# Undergraduate Lecture Notes in Physics

**Series Editors**

Neil Ashby, University of Colorado, Boulder, CO, USA

William Brantley, Department of Physics, Furman University, Greenville, SC, USA

Michael Fowler, Department of Physics, University of Virginia, Charlottesville, VA, USA

Morten Hjorth-Jensen, Department of Physics, University of Oslo, Oslo, Norway

Michael Inglis, Department of Physical Sciences, SUNY Suffolk County Community College, Selden, NY, USA

Barry Luokkala ⓘ, Department of Physics, Carnegie Mellon University, Pittsburgh, PA, USA

Undergraduate Lecture Notes in Physics (ULNP) publishes authoritative texts covering topics throughout pure and applied physics. Each title in the series is suitable as a basis for undergraduate instruction, typically containing practice problems, worked examples, chapter summaries, and suggestions for further reading.

ULNP titles must provide at least one of the following:

- An exceptionally clear and concise treatment of a standard undergraduate subject.
- A solid undergraduate-level introduction to a graduate, advanced, or non-standard subject.
- A novel perspective or an unusual approach to teaching a subject.

ULNP especially encourages new, original, and idiosyncratic approaches to physics teaching at the undergraduate level.

The purpose of ULNP is to provide intriguing, absorbing books that will continue to be the reader's preferred reference throughout their academic career.

More information about this series at https://link.springer.com/bookseries/8917

George Datseris · Ulrich Parlitz

# Nonlinear Dynamics

## A Concise Introduction Interlaced with Code

 Springer

George Datseris ⓘ
Max Planck Institute for Meteorology
Hamburg, Germany

Ulrich Parlitz ⓘ
Max Planck Institute for Dynamics
and Self-Organization
Göttingen, Niedersachsen, Germany

ISSN 2192-4791                    ISSN 2192-4805   (electronic)
Undergraduate Lecture Notes in Physics
ISBN 978-3-030-91031-0           ISBN 978-3-030-91032-7   (eBook)
https://doi.org/10.1007/978-3-030-91032-7

This Springer imprint is published by the registered company Springer Nature Switzerland AG
The registered company address is: Gewerbestrasse 11, 6330 Cham, Switzerland

# Preface

When we started writing this textbook, there were two goals we wanted to attain. The first goal was having a small, but up-to-date textbook, that can accompany a lecture (or a crash-course) on nonlinear dynamics with heavy focus on practical application. The target audiences therefore are on one hand students wanting to start working on applications of nonlinear dynamics, and on the other hand researchers from a different field that want to use methods from nonlinear dynamics in their research. Both groups require a concise, illustrative summary of the core concepts and a much larger focus on how to apply them to gain useful results.

The second goal for this book is to emphasize the role of computer code and explicitly include it in the presentation. As you will see, many (in fact, most) things in nonlinear dynamics are not treatable analytically. Therefore, most published papers in the field of nonlinear dynamics run plenty of computer code to produce their results. Unfortunately, even to this day, most of these papers do not publish their code. This means that reproducing a paper, or simply just implementing the proposed methods, is typically hard work, while sometimes even impossible. Textbooks do not fare much better either, with only a couple textbooks on the field discussing code. We find this separation between the scientific text and the computer code very unfitting for the field of nonlinear dynamics, especially given how easy it is to share and run code nowadays. We believe that the best way to solve the problem is teaching new generations to embrace code as part of the learning subject and as an integral part of science, and this is exactly what we are trying to do with this book.

## In a Nutshell

There are many textbooks on nonlinear dynamics, some of which are exceptionally good theoretical introductions to the subject. We think our approach will add something genuinely new to the field and bridge theory with real-world usage due to being:

- *practical.* We discuss quantities and analyses that are beneficial to compute in practice. Since just defining a quantity is typically not equivalent with computing it, we also provide algorithms and discuss limitations, difficulties, and pitfalls of these algorithms.
- *full of runnable code.* The algorithms and figures are accompanied with real, high-quality runnable code snippets *in-text*. Real code eliminates the inconsistencies of pseudo-code, as well as the hours (if not days) needed to implement it, while also enabling instant experimentation.
- *up-to-date.* Each chapter of the book typically highlights recent scientific progress on the subject of the chapter. In addition, the last chapters of the book are devoted to applications of nonlinear dynamics in topics relevant to the status quo, for example, epidemics and climate.
- *small.* This book is composed of 12 chapters, and we enforced ourselves that each chapter is small enough that it can be taught during a standard lecture week (we tested it!).

## Structure

The main text firstly explains the basic concepts conceptually but still accurately, without going too deep into the details. We then proceed to discuss algorithms, applications, and practical considerations. After that we (typically) present some runnable code that computes aspects of the taught concepts. A "Further reading" section at the end then provides historical overviews and relevant references as well as sources where one can go into more depth regarding the discussed concepts. Each chapter ends with selected exercises, some analytic while some applied, which aid understanding and sharpen practical skills.

## Usage in a Lecture

The precise size of the book was chosen on purpose so that each chapter spans a lecture week, and typical semesters span 12–13 weeks. Thus, it covers a full introductory course on nonlinear dynamics on an undergraduate or graduate program, depending on the students' background. We strongly recommend active teaching, where students are involved as much as possible in the lecture, e.g., by presenting multiple choice questions during the course. We also strongly recommend to run code live during the lecture, use the interactive applications we have developed in parallel with this book (see Sect. 1.3.1), and solve exemplary exercises. Pay special attention to the exercises that are oriented towards actively writing and using computer code. This also teaches being critical of computer results and being aware of common pitfalls.

While prior knowledge on calculus, linear algebra, probability, differential equations, and basic familiarity with programming are assumed throughout the book, its light mathematical tone requires only the basics. As a side note, we have purposely written the book in a casual tone, even adding jokes here and there. This is not only in line with our characters "in real life" but also because we believe it makes the book more approachable. We hope you enjoy it, and have fun with it (as much fun as one can have with a textbook anyways)!

Hamburg, Germany                                                        George Datseris
Göttingen, Germany                                                       Ulrich Parlitz
December 2021

# Acknowledgements

We would like to thank: Lukas Hupe and Jonas Isensee, two excellent students that served as teaching assistants for the courses that tested the book and some exercises; some expert reviewers that helped us ensure the rigor of some chapters: Michael Wilczek, Romain Veltz, Marc Timme, Thomas Lilienkamp, and Peter Ashwin; all contributors of the JuliaDynamics organization, especially those who have contributed to the DynamicalSystems.jl library. Ulrich Parlitz thanks H.D.I. Abarbanel, P. Bittihn, J. Bröcker, U. Feudel, R. Fleischmann, T. Geisel, F. Hegedűs, S. Herzog, H. Kantz, S. Klumpp, J. Kurths, L. Kocarev, W. Lauterborn, K. Lehnertz, C. Letellier, T. Lilienkamp, S. Luther, C. Masoller, R. Mettin, T. Penzel, A. Pikovsky, M. Rosenblum, A. Schlemmer, E. Schöll, H. Suetani, M. Timme, I. Tokuda, A. Witt, N. Wessel, F. Wörgötter, and many other (former) colleagues and students for inspiring discussions, very good cooperation and support.

George Datseris appreciated mentorship and support from Ragnar Fleischmann, Theo Geisel, Hauke Schmidt and Bjorn Stevens. And finally, we thank our friends and family for moral support during the development of the book and our colleagues for tolerating us being slow with other projects going on in parallel!

# Contents

# Chapter 1
# Dynamical Systems

**Abstract** This introductory chapter defines dynamical systems, stresses their wide spread applications, and introduces the concept of "deterministic chaos". Since computer simulations are valuable and even necessary for studying nonlinear dynamics we also show in this chapter examples of runnable code snippets that will be used throughout the book. We then look at fixed points, and when they are stable. To do so we employ linear stability analysis and subsequently discuss how volumes in the state space grow or shrink. We close by reviewing the Poincaré surface of section, a technique to convert a continuous system into discrete one, useful for visualizing higher-dimensional dynamics.

## 1.1 What Is a Dynamical System?

The term "dynamical system" can describe a wide range of processes and can be applied to seemingly all areas of science. Simply put, a dynamical system is a set of variables, or quantities, whose values change with time according to some predefined rules. These rules can have stochastic components, but in this book we will be considering systems without random parts, i.e., *deterministic dynamical systems*.

The variables that define the system are formally called the *state variables* of the system, and constitute the *state* or *state vector* $\mathbf{x} = (x_1, x_2, \ldots, x_D)$. For example, state variables can be the positions and velocities of planets moving in a gravitational field, the electric current and voltage of an electronic circuit, or temperature and humidity of a weather model, to name a few. The space that $\mathbf{x}$ occupies is called the *state space* or *phase space* $\mathcal{S}$ and has dimension[1] $D$, with $D$ the amount of variables that compose $\mathbf{x}$.

---

[1] Typically the state space $\mathcal{S}$ is an Euclidean space $\mathbb{R}^D$. But in general, it can be any arbitrary $D$-dimensional manifold. For example if some variables are by nature periodic, like the angle of a pendulum, the state space becomes toroidal.

© The Author(s), under exclusive license to Springer Nature Switzerland AG 2022
G. Datseris and U. Parlitz, *Nonlinear Dynamics*, Undergraduate Lecture Notes in Physics,
https://doi.org/10.1007/978-3-030-91032-7_1

Dynamical systems can be classified according to the nature of time. In this book we will be mainly considering two classes of dynamical systems, both having a continuous state space. This means that each component of $\mathbf{x}$ is a real number.[2] The first class is called *discrete* dynamical system

$$\mathbf{x}_{n+1} = f(\mathbf{x}_n) \tag{1.1}$$

with $f$ being the *dynamic rule* or *equations of motion* specifying the temporal evolution of the system. Here time is a discrete quantity, like steps, iterations, generations or other similar concepts. At each time step $f$ takes the current state and maps it to the next state: $x_1 = f(x_0)$, $x_2 = f(x_1)$, etc. Discrete systems are usually given by iterated maps as in (1.1), where the state $\mathbf{x}_n$ is plugged into $f$ to yield the state at the next step, $\mathbf{x}_{n+1}$. In our discussion here and throughout this book all elements of the vector-valued function $f$ are real. Any scenario where $f$ inputs and outputs complex values can be simply split into more variables containing the real and imaginary parts, respectively.

In *continuous* dynamical systems

$$\dot{\mathbf{x}} \equiv \frac{d\mathbf{x}}{dt} = f(\mathbf{x}) \tag{1.2}$$

time is a continuous quantity and $f$ takes the current state and returns the rate of change of the state. Continuous systems are defined by a set of coupled ordinary differential equations (ODEs). $f$ is also sometimes called *vector field* in this context. Equation (1.2) is a set of *autonomous* ordinary differential equations, i.e., $f$ does not explicitly depend on time $t$. This is the standard form of continuous dynamical systems. In general, non-autonomous systems can be written as autonomous systems with an additional state variable corresponding to time (see Chap. 9). Furthermore, (1.2) contains only first order time derivatives. All expressions with higher order derivatives (e.g., $\ddot{x} = ...$, obtained in Newtonian mechanics) can be re-written as first-order systems by introducing new variables as derivatives, e.g., $y = \dot{x}$. Notice the fundamental difference between continuous and discrete systems: in the latter $f$ provides the direct change, while in the former $f$ provides the rate of change.

### 1.1.1 Some Example Dynamical Systems

To put the abstract definition of dynamical systems into context, and also motivate them as tools for studying real world scenarios, let's look at some of the dynamical

---

[2] In Chap. 11, we will look at spatiotemporal systems, where each variable is a spatial field that is evolved in time.

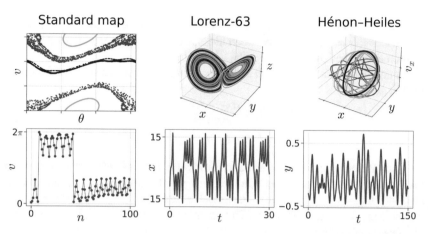

**Fig. 1.1** Top row: state space of example systems (for Hénon-Heiles only 3 of the 4 dimensions are shown). Three different initial conditions (different colors) are evolved and populate the state space. Bottom row: example timeseries of one of the variables of each system (only one initial condition is used). Each column is a different system, evolved using Code 1.1. Animations and interactive applications for such plots can be found online at animations/1/trajectory_evolution.

systems that we will be using in the subsequent chapters of this book. We visualize some of them in Fig. 1.1.

The simplest dynamical system, and arguably the most famous one, is the logistic map, which is a discrete system of only one variable defined as

$$x_{n+1} = r x_n (1 - x_n), \quad r \in [0, 4], \quad x \in [0, 1]. \tag{1.3}$$

This dynamical equation has been used in different contexts as an example or a prototype of a simple system that displays very rich dynamics nevertheless, with a prominent role in population dynamics. There it describes how a (normalized) population $x$ of, let's say, rabbits changes from generation to generation. At each new generation $n$ the number of rabbits increases because of $rx$, with $r$ a growth factor (which is of course the available amount of carrots). But there is a catch; if the rabbit population becomes too large, there aren't enough carrots to feed all of them! This is what the factor $(1 - x)$ represents. The higher the population $x_n$ at generation $n$, the more penalty to the growth rate.

Another well-studied prototypical two dimensional discrete system is the standard map, that is given by

$$\begin{aligned} \theta_{n+1} &= \theta_n + v_n + k \sin(\theta_n) \\ v_{n+1} &= v_n + k \sin(\theta_n) \end{aligned} \tag{1.4}$$

with $\theta$ the angle and $v$ the velocity of an oscillating pendulum on a tabletop (gravity-free). The pendulum is periodically kicked at every time unit by a force of strength $k$ ($k = 1$ unless noted otherwise), which has a constant direction (thus the term sin). You may wonder why this is a discrete system when clearly an oscillating pendulum is operating in continuous time. The reason is simple: since the kick is applied only at each step $n$, and the motion between kicks is free motion, we only need to record the momentum and angle values at the time of the kick.

Yet another famous dynamical system, continuous in this case, is the Lorenz-63 system, given by

$$\begin{aligned}
\dot{x} &= \sigma(y - x) \\
\dot{y} &= -xz + \rho x - y \\
\dot{z} &= xy - \beta z.
\end{aligned} \tag{1.5}$$

It is a simplified model for atmospheric convection, representing a two-dimensional fluid layer. $x$ is the rate of convection and $y$ and $z$ are the temperature variations in the horizontal and vertical direction of the fluid, respectively. The parameters $\rho, \beta, \sigma$ are related to the properties of the fluid like its viscosity (we use $\rho = 28, \beta = 8/3, \sigma = 10$ unless noted otherwise).

The last system that we list here is the Hénon-Heiles system

$$\begin{aligned}
\dot{x} &= v_x, & \dot{v}_x &= -x - 2xy \\
\dot{y} &= v_y, & \dot{v}_y &= -y - (x^2 - y^2).
\end{aligned} \tag{1.6}$$

This is a simplification of the motion of a star (represented by a point particle with positions $x, y$ and velocities $v_x, v_y$) around a galactic center (positioned at $(0, 0)$). Although we will discuss how to simulate dynamical systems in detail in Sect. 1.3, let's see how these systems "look like" when evolved in time. In Fig. 1.1 we show both the state space of some systems (populated by evolving three different initial conditions, each with a different color) as well as an example timeseries of one of the variables.

It may seem that these examples were handpicked to be as diverse as possible. This is not at all the case and we were honest with you in the introduction. The framework of dynamical systems applies in seemingly arbitrary parts of reality.

### 1.1.2 Trajectories, Flows, Uniqueness and Invariance

A *trajectory* (commonly also called an *orbit*) represents the evolution from an initial condition $\mathbf{x}_0 = \mathbf{x}(t = 0)$ in the state space. *Fixed points*, defined by $f(\mathbf{x}^*) = \mathbf{x}^*$ (discrete dynamics) or $f(\mathbf{x}^*) = 0$ (continuous dynamics), are technically trajectories that consist of a single point, i.e., a state that never changes in time. Another special kind of trajectories are periodic ones, which for discrete systems are a finite number of $N$ points satisfying $\mathbf{x}_{n+N} = \mathbf{x}_n$, and for continuous systems are closed curves in the state space satisfying $\mathbf{x}(t + T) = \mathbf{x}(t)$. Putting these two cases aside, for discrete systems a trajectory consists of an infinite, but ordered, number of points and for continuous systems trajectories are curves in the state space that do not close.

In most cases (and for all systems we will consider in this book), the Picard-Lindelöf theorem states that for each initial condition $\mathbf{x}_0$ a unique solution $\mathbf{x}(t)$

of the ODE(s) (1.2) exists for $t \in (-\infty, \infty)$ if the vector field $f$ fulfills a mild smoothness condition, called Lipschitz continuity[3] which is met by most systems of interest. Only in cases where the solution diverges to infinity in finite time (that will not be considered in this book) $\|\mathbf{x}(t \to T_\pm)\| \to \infty$, the existence limits are finite, $t \in (T_-, T_+)$. The uniqueness property implies that any trajectory $\{\mathbf{x}(t) \in S : -\infty < t < \infty\}$ never intersects itself (except for fixed points or periodic orbits) or trajectories resulting from other initial conditions, i.e., the time evolution is unique.

For discrete systems forward-time uniqueness is always guaranteed for any $f$, regardless of continuity, simply from the definition of what a function is. To have the same feature backward in time $f$ has to be invertible, in contrast to continuous systems where this is not required.

In simulations (or with measured data) trajectories of continuous systems are sampled at discrete time points which creates a set of states $A = \{\mathbf{x}_0, \mathbf{x}_1, \ldots, \mathbf{x}_N\}$. If only a single component of the state vectors $\mathbf{x}_n$ or a scalar function of the states is considered this provides a *uni-variate timeseries* or just *timeseries*. Measuring several observables simultaneously yields a *multivariate timeseries*.

Another concept which is useful for describing the temporal evolution is the *flow* $\Phi^t(\mathbf{x})$, a function mapping any state $\mathbf{x}$ to its future ($t > 0$) or past ($t < 0$) image $\mathbf{x}(t) = \Phi^t(\mathbf{x})$ (here $t$ could also be $n$, i.e., the same concept holds for both continuous or discrete systems). Note that $\Phi^0(\mathbf{x}) = \mathbf{x}$ and $\Phi^{t+s}(\mathbf{x}) = \Phi^t(\Phi^s(\mathbf{x})) = \Phi^s(\Phi^t(\mathbf{x}))$. An *invariant set* is by definition a set $A$ which satisfies $\Phi^t(A) = A$, $\forall t$. This means that the dynamics maps the set $A$ to itself, i.e., it is "invariant" under the flow. Invariant sets play a crucial role in the theoretical foundation of dynamical systems.

### 1.1.3  Notation

In the remainder of this book we will typically use capital letters $A, X, Y, Z, \Omega$ to denote sets (sampled trajectories are ordered sets) and matrices, bold lowercase upright letters $\mathbf{a}, \mathbf{x}, \mathbf{y}, \mathbf{z}, \omega$ to notate vectors, and lowercase letters $a, x, y, z, \omega$ to notate timeseries, variables or constants. A few exceptions exist, but we believe there will be no ambiguities due to the context. The symbols $f, D, p, P$ are used exclusively to represent: the dynamic rule (also called equations of motion or vector field), the dimensionality of the state space or set at hand, an unnamed parameter of the system, and a probability, respectively. Accessing sets or timeseries at a specific *index* is done either with subscripts $x_n$ or with brackets $x[n]$ ($n$ is always integer here). The syntax $x(t)$ means "the value of $x$ at time $t$" in the case where $t$ is continuous time, or it means that $x$ is a function of $t$. The symbol $\|\mathbf{x}\|$ means the norm of $\mathbf{x}$, which depends on the metric of $S$, but most often is the Euclidean norm (or metric). Set elements are enclosed in brackets $\{\}$.

---

[3] A function $f : S \to S$ is called Lipschitz continuous if there exists a real constant $k \geq 0$ so that $\|f(\mathbf{x}) - f(\mathbf{y})\| \leq k\|\mathbf{x} - \mathbf{y}\|$ for all $\mathbf{x}, \mathbf{y} \in S$. Every continuously differentiable function is Lipschitz continuous.

### *1.1.4  Nonlinearity*

A linear equation (or system) $f$ by definition satisfies the superposition principle $f(\alpha\mathbf{x} + \beta\mathbf{y}) = \alpha f(\mathbf{x}) + \beta f(\mathbf{y})$, $\alpha, \beta \in \mathbb{R}$ and thus any two valid solutions can be added to generate a new solution. Furthermore, the norm of the state $\mathbf{x}$ is irrelevant, as a solution can be given for any scale by changing $\alpha$, and thus only the orientation of $\mathbf{x}$ matters. Because of these two properties linear dynamical systems can be solved in closed form and do not exhibit any complex dynamics.

You may have noticed that the rules $f$ that we have described in Sect. 1.1.1 are *nonlinear* functions of the state $\mathbf{x}$. In contrast to linear systems, nonlinear systems must be *considered as a whole* instead of being able to be split into smaller and simpler parts, and on a *scale-by-scale basis*, as different scales are dominated by different dynamics within the same system. Therefore, nonlinear systems display a plethora of amazing features and dominate in almost all natural processes around us.

## 1.2  Poor Man's Definition of Deterministic Chaos

The systems we are considering in this book are nonlinear and *deterministic*. Given an initial condition and the rule $f$ we are in theory able to tell everything about the future of the system by evolving it forward in time. But one must not confuse deterministic systems with simple or easy-to-predict behaviour. While sometimes indeed deterministic dynamics will lead to simple, periodic behavior, this is not always the case as illustrated in the bottom row of Fig. 1.1.

Interestingly, the timeseries shown *never repeat themselves* in a periodic manner,[4] even though the systems are deterministic! On the other hand, the timeseries aren't random either, as they seem to be composed of easily distinguishable patterns. For example, in the case of the standard map, one sees that sequences of points form triangles, and they are either near $2\pi$ pointing down or near 0 pointing up. Similar patterns exist in the Lorenz-63 and Hénon-Heiles model, where there are oscillations present, but they are non-periodic. It feels intuitive (and is also precisely true) that these patterns have a dynamic origin in the rule of the system $f$. Their exact sequence (i.e., when will the triangles in the standard map switch from up to down) may appear random, however, and sometimes may indeed be statistically equivalent to random processes.

This is called *deterministic chaos*: when the evolution of a system in time is non-periodic, apparently irregular, difficult to predict, but still deterministic and full of patterns. These patterns result in *structure in the state space*, as is seen clearly in the top row of Fig. 1.1. The sets that are the result of evolving a chaotic system (or more precisely, an initial condition that leads to chaotic dynamics in that system) will be called *chaotic sets* in this book. The fact that these sets have structure in the state space is the central property that distinguishes deterministic chaos from pure randomness: even if the sequence (timeseries) appears random, in the state space

---

[4] This might not be obvious from the small time window shown in the plot, so for now you'll have to trust us or simulate the systems yourself based on the tools provided in Sect. 1.3.1.

there is structure (of various forms of complexity). We will discuss deterministic chaos in more detail in Chap. 3.


## 1.3 Computer Code for Nonlinear Dynamics


As nonlinear dynamical systems are rarely treatable fully analytically, one always needs a computer to study them in detail or to analyze measured timeseries. Just a computer though is not enough, but one also needs code to run! This is the reason why we decided to write this book so that its pages are interlaced with real, runnable computer code. Having a "ready-to-go" piece of code also enables instant experimentation on the side of the reader, which we believe to be of utmost importance.

To write real code, we chose the Julia programming language because we believe it is highly suitable language for scientific code and even more so for the context of this book. Its simple syntax allows us to write code that corresponds line-by-line to the algorithms that we describe in text. Furthermore, Julia contains an entire software organization for dynamical systems, JuliaDynamics whose software we use throughout this book. The main software we'll be using is DynamicalSystems.jl, a software library of algorithms for dynamical systems and nonlinear dynamics, and InteractiveDynamics.jl, which provides interactive graphical applications suitable for the classroom. To demonstrate, let's introduce an example code snippet in Code 1.1, similar to the code that we will be showing in the rest of the book. We think that the code is intuitive even for readers unfamiliar with Julia.

**Code 1.1** Example code defining the Lorenz-63 system, Eq. (1.5) in Julia, and obtaining a trajectory for it using DynamicalSystems.jl.

```julia
using DynamicalSystems # load the library

function lorenz_rule(u, p, t) # the dynamics as a function
    σ, ρ, β = p
    x, y, z = u
    dx = σ*(y - x)
    dy = x*(ρ - z) - y
    dz = x*y - β*z
    return SVector(dx, dy, dz) # Static Vector
end

p  = [10.0, 28.0, 8/3] # parameters: σ, ρ, β
u₀ = [0.0, 10.0, 0.0]  # initial condition
# create an instance of a `DynamicalSystem`
lorenz = ContinuousDynamicalSystem(lorenz_rule, u₀, p)

T  = 100.0 # total time
Δt = 0.01  # sampling time
A  = trajectory(lorenz, T; Δt)
```

Code is presented in mono-spaced font with a light purple background, e.g., `example` . Code 1.1 is an important example, because it shows how one defines a "dynamical system" in Julia using DynamicalSystems.jl.[5] Once such a dynamical system is defined in code, several things become possible through DynamicalSystems.jl. For example, in Code 1.1 we use the function `trajectory` to evolve the initial condition of a `DynamicalSystem` in time, by solving the ODEs through DifferentialEquations.jl.

### 1.3.1  Associated Repository: Tutorials, Exercise Data, Apps

Julia, DynamicalSystems.jl, and in fact anything else we will use in the code snippets are very well documented online. As this is a textbook about nonlinear dynamics, and not programming, we will not be teaching Julia here. Besides, even though we use Julia here, everything we will be presenting in the code snippets could in principle be written in other languages as well.

Notice however that Julia has an in-built help system and thus most snippets can be understood without consulting online documentations. Simply put, in the Julia console you can type question mark and then the name of the function that you want to know more about, e.g., `?trajectory` . Julia will then display the documentation of this function, which contains detailed information about exactly what the function does and how you can use it. Another reason not to explain code line-by-line here is because programming languages and packages get updated much, much faster than books do, which runs the risk of the explanations in a book becoming obsolete.

There is an online repository associated with this book, https://github.com/JuliaDynamics/NonlinearDynamicsTextbook. There we have collected a number of tutorials and workshops that teach core Julia as well as major packages. This repository also contains all code snippets and all code that produces the figures that make up this book, and there we can update code more regularly than in the book. It also contains the exact package versions used to create the figures, establishing reproducibility. The same repository contains the datasets that are used in the exercises of every chapter, as well as multiple choice questions that can be used during lectures (e.g., via an online polling service), but also in exams. Perhaps the most important thing in this repository are scripts that launch interactive, GUI-based applications (apps) that elucidate core concepts. These are present in the folder animations and are linked throughout the book. For convenience, a recorded video is also provided for each app.

---

[5] In the following chapters we will not be defining any new dynamical systems in code, but rather use predefined ones from the `Systems` submodule.

### *1.3.2 A Notorious Trap*

Almost every algorithm that we will be presenting in this book has a high quality performant implementation in DynamicalSystems.jl. In addition, we strongly believe that code should always be shared and embraced in scientific work in nonlinear dynamics. New papers should be published with the code used to create them, establishing reproducibility, while new algorithms should be published with a code implementation (hopefully directly in DynamicalSystems.jl). However, one must be aware that having code does not replace having knowledge. Having a pre-made implementation of an algorithm can lead into the trap of "just using it", without really knowing what it does or what it means. Thus, one must always be vigilant, and only use the code once there is understanding. We recommend students to attempt to write their own versions of the algorithms we describe here, and later compare with DynamicalSystems.jl implementations. Since DynamicalSystems.jl is open source, it allows one to look inside every nook and cranny of an algorithm implementation. Another benefit of using Julia for this book is how trivial it is to access the source code of any function. Simply preface any function call with `@edit`, e.g. `@edit trajectory(...)`, and this will bring you to the source code of that function.

## 1.4 The Jacobian, Linearized Dynamics and Stability

Previously we've mentioned *fixed points*, that satisfy $f(\mathbf{x}^*) = 0$ for continuous systems and $f(\mathbf{x}^*) = \mathbf{x}^*$ for discrete systems. In the rest of this section we care to answer one simple, but important question: when is a fixed point of a dynamical system "stable"? We will answer this question by coming up with an intuitive definition of stability, in terms of what happens infinitesimally close around a state space point $\mathbf{x}$ as time progresses.

Let's take a fixed point $\mathbf{x}^*$ and perturb it by an infinitesimal amount $\mathbf{y}$. We are interested in the dynamics of $\mathbf{x} = \mathbf{x}^* + \mathbf{y}$ and whether $\mathbf{x}$ will go further away or towards $\mathbf{x}^*$ as time progresses. For simplicity let's see what is happening in the continuous time case, and because $\mathbf{y}$ is "very small", we can linearize the dynamics with respect to $\mathbf{x}^*$, using a Taylor expansion

$$\dot{\mathbf{x}} = \dot{\mathbf{y}} = f(\mathbf{x}^* + \mathbf{y}) = f(\mathbf{x}^*) + J_f(\mathbf{x}^*) \cdot \mathbf{y} + \cdots \tag{1.7}$$

where $J_f(\mathbf{x}^*)$ stands for the Jacobian matrix of $f$ at $\mathbf{x}^*$ with elements

$$J_f(\mathbf{x}^*)_{ij} = \left. \frac{\partial f_i(\mathbf{x})}{\partial \mathbf{x}_j} \right|_{\mathbf{x}=\mathbf{x}^*} \quad \text{or} \quad J_f(\mathbf{x}^*) = \begin{bmatrix} \dfrac{\partial f_1}{\partial x_1} & \cdots & \dfrac{\partial f_1}{\partial x_D} \\ \vdots & \ddots & \vdots \\ \dfrac{\partial f_D}{\partial x_1} & \cdots & \dfrac{\partial f_D}{\partial x_D} \end{bmatrix}_{\mathbf{x}=\mathbf{x}^*} \tag{1.8}$$

and we have also stopped expanding terms beyond first order (the Jacobian represents the first order).

Since $f(\mathbf{x}^*) = 0$ by definition, the dynamics for small $\mathbf{y}$ and for short times is approximated by the behaviour of $\dot{\mathbf{y}} = J_f(\mathbf{x}^*) \cdot \mathbf{y}$, which is a *linear* dynamical system. This equation is also called the *linearized* dynamics, or the *tangent* dynamics, or the dynamics in the *tangent space*. Notice that because $\dot{\mathbf{y}} = J_f(\mathbf{x}^*) \cdot \mathbf{y}$ is a linear system, the actual size of $\mathbf{y}$ does not matter for its evolution. But, one here must clearly distinguish the case of evolving $\mathbf{y}$ using the linear dynamics or using the full nonlinear dynamics. If the perturbation $\mathbf{y}$ is very small, then linear and nonlinear evolution are in fact approximately the same (see Hartman-Grobman theorem below). But as $\mathbf{y}$ increases in size the nonlinear effects become increasingly more important and evolving $\mathbf{y}$ with the linearized dynamics or the full nonlinear dynamics is no longer equivalent.

In any case, since in the current discussion we care about stability very close to $\mathbf{x}^*$ (and thus $\mathbf{y}$ is indeed small), the linearized dynamics is sufficient to describe the temporal evolution near $\mathbf{x}^*$. The eigenvalues of the Jacobian $J_f(\mathbf{x}^*) : \mu_1, \ldots, \mu_D$, characterize the stability of this linear dynamics completely. Recall that the linear 1D continuous equation $\dot{z} = az$ has the solution $z(t) = z_0 \exp(at)$. With this in mind, we can explicitly write down the solution for $\mathbf{y}(t)$, by expressing $\mathbf{y}_0$ in the basis of the Jacobian eigenvectors[6]

$$\mathbf{y}(t) = \sum_{i=1}^{D} c_i \mathbf{e}_i \exp(\mu_i t) \tag{1.9}$$

with $c_i$ the coefficients and $(\mathbf{e}_i, \mu_i)$ the Jacobian eigenvector-eigenvalues pairs. Since $J$ is a real matrix, $\mu_i$ are either real or complex-conjugate pairs. If they are real, then $\mu_i$ is the rate of increase (or decrease if negative) of the perturbation along the $i$-th eigenvector of $J_f$, as it becomes clear from (1.9). If some $\mu_i$ are complex-conjugate pairs, then the imaginary part corresponds to a rotation in the plane spanned by the two corresponding eigenvectors, while the real part as before corresponds to the rate of increase or decrease. This approach of looking at the eigenvalues of the Jacobian is called *linear stability analysis*.

The fixed point $\mathbf{x}^*$ is *stable* if the state $\mathbf{x}$ comes closer and closer to it during its temporal evolution. This happens if $\forall i : \text{Re}(\mu_i) < 0$, i.e., $\mathbf{x}^*$ attracts $\mathbf{x}$ towards it in all directions. If, however, there is at least one $i$ so that $\text{Re}(\mu_i) > 0$, then there is at

---

[6] Equation (1.9) assumes the Jacobian has full rank. The modification when this is not the case is straightforward.

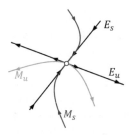

**Fig. 1.2** Stable and unstable subspaces $E_s$ and $E_u$, respectively, of a saddle point in 2D. At the fixed point these subspaces are tangential to the stable and unstable manifolds, $M_s$ and $M_u$, emerging from the fixed point

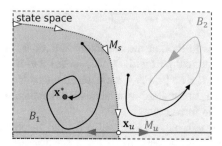

**Fig. 1.3** Two attractors in the state space of a continuous dynamical system: a stable fixed point and stable limit cycle (see Chap. 2), and their basins of attraction $B_1$, $B_2$ in purple/cyan color. The stable/unstable manifolds of a nearby unstable fixed point $\mathbf{x}_u$ are also plotted, which often serve as boundaries of basins of attraction. See also Fig. 5.4

least one direction in the state space that $\mathbf{x}$ will move further and further away from $\mathbf{x}^*$. In this case $\mathbf{x}^*$ is an *unstable* fixed point.

Stable and unstable directions are given by the corresponding eigenvectors of the Jacobian. These are spanning stable and unstable linear subspaces $E_s$ and $E_u$, respectively, at the fixed point as illustrated in Fig. 1.2. Asymptotically close to the fixed point, the evolution of a state in the linear subspaces stays in the subspaces. But as we move further away from the fixed point, the stable and unstable directions are given by stable and unstable manifolds $M_s$ and $M_u$ shown in Fig. 1.2. These manifolds are invariant with respect to the flow, i.e., any state in $M_u$ moves away from the fixed point but stays in $M_u$. And states in $M_s$ will approach $\mathbf{x}^*$ with $\mathbf{x}(t) \in M_s \; \forall t$. At the fixed point these manifolds are tangential to the corresponding eigenspaces of the linearized dynamics. The manifolds $M$ are important elements of the dynamics as they organize the state space, acting like a "skeleton" guiding the flow. Also, they often define the boundaries of basins of attraction of different attractors in the system, see Sect. 1.5 and see Fig. 1.3 for more.

What happens at the equals case, where $\text{Re}(\mu_i) = 0$? This case seems stable, since $\mathbf{x}$ is not diverging away from $\mathbf{x}^*$ in the $i$-direction. Unfortunately, we cannot know. The Hartman-Grobman theorem states that for a small neighborhood around a fixed point $\mathbf{x}^*$, the linearized dynamics is topologically conjugate (i.e., equivalent) to the nonlinear dynamics, if and only if the Jacobian $J_f(\mathbf{x}^*)$ has no eigenvalue with zero real part. Thus, if there is any $\text{Re}(\mu_i) = 0$ we cannot use linear stability analysis to make any claims for the stability of the nonlinear system because in this case the stability of the nonlinear system at $\mathbf{x}^*$ depends on higher-order terms of the Taylor expansion (1.7).

The situation for discrete dynamical systems is quite similar. The only difference now is that the eigenvalues of $J_f$ do not give the rate of change, but the direct change instead. The equation of the linearized dynamics here becomes $\mathbf{y}_{n+1} = J_f(\mathbf{x}^*) \cdot \mathbf{y}_n$. By keeping in mind that the linear 1D discrete system $z_{n+1} = a z_n$ has the explicit solution is $z_n = z_0 a^n$, we can again make a solution for $\mathbf{y}_n$ by expanding on the eigenvectors of the Jacobian

$$\mathbf{y}_n = \sum_{i=1}^{D} c_i \mathbf{e}_i \mu_i^n. \tag{1.10}$$

Therefore, now one has to look at the absolute value of $\mu_i$ and compare it with 1 instead of comparing the real part with 0. This means that if $\forall i : |\mu_i| < 1$ we have a stable fixed point but if $\exists i : |\mu_i| > 1$ we have an unstable fixed point. If there is any eigenvalue $|\mu_i| = 1$, linear stability analysis is not sufficient to make a statement about the stability of the fixed point.

In the following we will call an eigenvalue (of the Jacobian matrix) *unstable* if $\text{Re}(\mu_i) > 0$ for continuous systems, or $|\mu_i| > 1$ for discrete. The notion of *stable* eigenvalue will be used for $\text{Re}(\mu_i) < 0$ (continuous) or $|\mu_i| < 1$ (discrete). Fixed points that have both stable and unstable eigenvalues are called *saddle* points or *hyperbolic* points.

## 1.5   Dissipation, Attractors, and Conservative Systems

For discrete systems the linearized dynamics at the fixed point is contracting if the magnitude of the determinant of its Jacobian $|\det(J_f)| = \left|\prod_i \mu_i\right|$ is smaller than one. Here contracting means that the volume of an infinitesimal ball of initial conditions around the fixed point is shrinking in time. Since we are considering the volume (or area in 2D) this could happen even for unstable fixed points if the contracting dynamics in some directions is faster than the expansion in the unstable directions. For continuous systems, the volume near a fixed point is contracting if the trace of the Jacobian $\text{trace}(J_f) = \text{div}(f) = \sum_i \mu_i$ is negative. Since $f$ is real, the eigenvalues of the Jacobian matrix can only be real or complex-conjugate pairs, thus both $\det(J_f)$ as well as $\text{trace}(J_f)$ are always real.

This characterization of the evolution of small volumes holds not only for fixed points, but in the vicinity of any state $\mathbf{x}$. A system is called *dissipative* if for almost all[7] initial values $\mathbf{x}$ the volume $v$ of an infinitesimally small ball of initial conditions following the trajectory starting at $\mathbf{x}$ will converge to zero as $t \to \infty$. For discrete sys-

---

[7] Exceptions are, e.g., some unstable fixed points.

tems this is the case if the sequence $v_{n+1} = |\det(J_f(\mathbf{x}_n))| \cdot v_n$ (with $v_0 = 1$) converges to zero. With continuous systems the evolution of infinitesimal volumes in state space is given by[8] $\dot{v} = \text{trace}(J_f(\mathbf{x}(t)) \cdot v$ (with $v(0) = 1$) and dissipative dynamics is characterized by $\lim_{t \to \infty} v(t) = 0$. It may happen that the trajectory passes some regions in state space where the ball is actually expanded (because locally $|\det(J_f(\mathbf{x}))| > 1$ or $\text{trace}(J_f(\mathbf{x})) > 0$), but at the end "contraction has to win" once averaged over the trajectory. Exactly how such small volumes expand or contract will be discussed in more detail in Sect. 3.2.2.

The state space of dissipative systems is populated by *attractors*. These are invariant subsets of the state space that attract nearby initial conditions towards them. Initial conditions starting already in an attractor stay within it forever, both forwards and backwards in time. A stable fixed point is an attractor, for example, and another one is a stable limit cycle (an isolated periodic orbit that attracts nearby initial conditions, see Chap. 2). Another important concept is the *basin of attraction* of an attractor which is the set of initial conditions in state space from where you reach the attractor (or "fall into the attractor", hence the word "basin") when evolved forwards in time. The set of all initial conditions $\mathbf{x}(0) \in B$ that satisfy $\lim_{t \to \infty} ||\mathbf{x}(t) - A|| \to 0$ with $A$ the attractor set, are forming the basin of attraction $B$ of $A$ (here we exploit the notation $||\mathbf{x}(t) - A||$ to denote the distance of a point $\mathbf{x}$ from the point $\mathbf{a} \in A$ which is closest to $\mathbf{x}$). An illustration of attractors and basins of attraction is given in Fig. 1.3.

Since the dynamics is dissipative, all attractors are by definition sets of zero volume in the state space. Although it may seem counter-intuitive at first, some attractors can even be *chaotic* (also called *strange*), which means that they are complicated, typically fractal objects (see Chap. 5 for fractals). Chaotic attractors also have zero volume, even though they have the property that some specific directions of the state space are expanded instead of contracted. The flow of the Lorenz-63 system, shown in Fig. 1.1 (middle column), is following a chaotic attractor that looks like a butterfly.

An important class of systems are those where $|\det(J)| = 1$ for discrete or $\text{trace}(J) = 0$ for continuous systems *everywhere* in state space. In other words, contraction and expansion rates exactly balance each other. This class is called *conservative dynamical systems*[9] and includes Hamiltonian systems known from classical mechanics, like the standard map and the Hénon-Heiles system presented in Sect. 1.1.1. We will discuss (Hamiltonian) conservative system in more detail in Chap. 8.

In both conservative and dissipative systems, the actual type of asymptotic motion may depend on the initial condition. For example, in dissipative systems more than one attractor may exist in the state space and different initial conditions can converge to different attractors like in Fig. 1.3. This is called *multistability* or *coexisting attractors*. Similarly, for conservative systems it is often the case that some initial

---

[8] A derivation of this differential equation is given in Appendix A.

[9] Technically, each initial condition of a dynamical system corresponds to a dissipative or conservative evolution (or divergence to infinity), and different types can coexist in the same system. But in the majority of cases all state space shares the same type.

conditions lead to chaotic time evolution and some others to periodic or quasiperiodic (see Sect. 2.4 for quasiperiodicity). This situation is called *mixed state space*.

### 1.5.1 Conserved Quantities

Sometimes, the rule $f$ leads to some conserved quantities, and this can happen for both dissipative or conservative systems. These quantities are scalar functions of the state, e.g., $g(\mathbf{x})$ which do not change when $\mathbf{x}$ is evolved in time, i.e., $\dot{g} = 0$ or $g_{n+1} = g_n$. An example are Hamiltonian systems, which by definition are given by

$$\dot{q} = \partial H / \partial v, \quad \dot{v} = -\partial H / \partial q, \quad H = H(q, v) \tag{1.11}$$

with $H$ the Hamiltonian and $q, v$ the coordinate and momentum conjugate pair, respectively.[10] Provided that $H$ does not explicitly depend on time, using the differentiation chain rule one can show that $\dot{H} = 0$, i.e., the Hamiltonian (which corresponds to the energy of the classical system) is a conserved quantity whose value depends only on the initial condition $(q_0, v_0)$.

One way of finding such conserved quantities is to eliminate time dependence, e.g., trying to combine $dx/dt$ and $dy/dt$ into $dx/dy$. Using separation of variables, you can sometimes integrate the result to find some function $g(x, y)$ which has zero time derivative. Identifying conserved quantities is practically useful, because it is guaranteed that the time evolution of a trajectory must lie on the manifold defined by $g(\mathbf{x}) = \text{const.} = g(\mathbf{x}_0)$.

## 1.6 Poincaré Surface of Section and Poincaré Map

The Poincaré surface of section (PSOS) is a powerful tool that can "transform" a $D$-dimensional continuous dynamical system into a $(D - 1)$-dimensional discrete one described by the resulting *Poincaré map*. It is quite simple how it works: first one defines a $D - 1$ dimensional hyperplane in the state space of the continuous system (any arbitrary manifold of dimension $D - 1$ may be defined instead of a hyperplane, but here we keep things simple for illustration). The hyperplane must be appropriately chosen so that when a trajectory is evolved in the state space, it crosses the plane transversely, as shown in Fig. 1.4. One then records these points where the trajectory intersects the hyperplane, ensuring that only crossings in one direction through the hyperplane are recorded. Notice that to have a proper PSOS, the trajectory should cross the hyperplane again in its future and its past. Any point in the PSOS is thus in both directions in time mapped to the PSOS again and this operation

---

[10] In the literature the symbol $p$ is used to represent momentum, instead of $v$. However in our book $p$ is a reserved symbol for an unnamed system parameter.

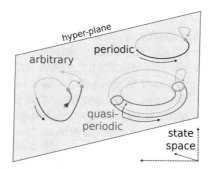

**Fig. 1.4** Demonstration of the Poincaré surface of section and map for various trajectories. Arrows indicate the direction of the trajectory. See Sect. 2.4 for quasiperiodic trajectories. Cyan points denote the actual section, while cyan dashed arrows illustrate the Poincaré map. Notice that only crossings in one direction are considered

is called the *Poincaré map* (or its inverse in case of past evolution). The Poincaré map is a full representation of the continuous dynamics due to the uniqueness and smoothness of solutions of ODEs. The Poincaré section and map are implemented in DynamicalSystems.jl as `poincaresos` and `poincaremap`. An interactive app showing a Poincaré section of the orbits of the Hénon-Heiles system from Fig. 1.1 can be found at animations/1/poincare.

Why are the PSOS and the Poincaré map such useful tools? Besides transforming continuous systems into discrete ones (by taking the sequence of crossings), it comes with an additional benefit of reducing the dimension of the crossed object by one. For example, a periodic orbit has dimension 1, since it's a closed loop. The periodic orbit will only cross the hyperplane at a finite number of points, and then simply pass through these points again and again. But any finite set of points has dimension 0! The same dimension reduction happens for the PSOS of a two-dimensional object. In Fig. 1.4 we show the section of a quasiperiodic trajectory on a 2-torus (think of it as the surface of a donut). If you cut it with a plane, what will remain on the plane will be a circle, which has dimension 1, one less than the 2-torus. The same process happens for objects of any dimensionality. In fact, in Chap. 5 we will see that some chaotic sets (like the attractor of the Lorenz-63 system) have some dimension $\Delta$ that is non-integer. The resulting PSOS of such a set has dimension $\Delta - 1$, which is also non-integer.

## Further Reading

A motivating introduction to nonlinear dynamics and chaos is a common science book by Gleick, *Chaos: Making a new science* [1]. The history behind some of the chapters of our book is presented there in an elegant and captivating manner. We strongly recommend this book to anyone interested in chaos from a more historic point of view. Reference [2] by Holmes contains a detailed historic overview of the mathematical origins of (nonlinear) dynamical systems theory, with plethora of further references. Diacu [3] provides another historical overview of the origins of chaos theory, this one focusing more on the originators and their encounters.

*Nonlinear dynamics and chaos* by Strogatz [4] is a well known textbook that provides a light introduction to dynamical systems, and in addition can be consulted by students who may not have a strong mathematics background. More specifically for the present chapter, [4] discusses what happens in the case when the Hartman-Grobman theorem (discussed in Sect. 1.4) fails, and the stability of a fixed point cannot be known from the linearized terms. Then nonlinear terms can be used to make statements about stability. An alternative well known textbook, that also expands in detail on introductory concepts, is *An Exploration of Dynamical Systems and Chaos* by Argyris, Faust, Haase and Friedrich [5]. See also the two books by Arnold, *Ordinary differential equations* [352] and *Geometrical methods in the Theory of Ordinary Differential Equations* [353] for fundamental concepts in differential equations and geometry of the state space.

The four systems that we introduced in Sect. 1.1.1 have had major influence on the development of nonlinear dynamics. May popularized the logistic map [6], Chirikov introduced the standard map [7], Lorenz the homonymous system [8], and Hénon-Heiles their conservative system [9]. These papers are certainly worth having a look.

## Selected Exercises

1.1  Find the fixed points of the logistic map, the standard map and the Hénon-Heiles system from Sect. 1.1.1 and discuss their stability.

1.2  Find period-2 orbits of the standard map. *Hint: solve* $\mathbf{x}_{n+2} = \mathbf{x}_n$.

1.3  For which parameter values is the Hénon map, $x_{n+1} = 1 - ax_n^2 + by_n$, $y_{n+1} = x_n$, a conservative system?

1.4  Prove that the Hénon-Heiles system (1.6) is conservative. Derive a condition for the parameters of the Lorenz-63 model (1.5) which has to be fulfilled such that this system is dissipative.

1.5  In a computer define and evolve the following dynamical system

$$\dot{x} = \mu x - yz, \quad \dot{y} = -vy + xz, \quad \dot{z} = \gamma - z + xy$$

(by default $\mu = 0.119$, $v = 0.1$, $\gamma = 0.9$ and initial condition $(3, 3, 3)$) called Gissinger system [10], which models chaotic reversals of Earth's magnetic field. Animate the evolution of the three dimensional trajectory for various values of $\mu$. *Hint: writing your own ODE solver is probably too much; instead pick an existing one from your programming language of choice. You may compare with the solution you'd get by using DynamicalSystems.jl and Code 1.1.*

1.6  Evolve a grid of initial conditions of the standard map, (1.4), each for 1000 steps. Scatter-plot all trajectories of all initial conditions with different colors.

Repeat for various $k$ values in different plots. These plots will highlight that the standard map is a system with mixed state space, which we will discuss more in Sect. 8.2.2.

1.7 Show that for Hamiltonian systems the Hamiltonian is a conserved quantity, $\dot{H} = 0$. Apply this knowledge to the single pendulum:

$$\dot{\theta} = v, \quad \dot{v} = -p\sin(\theta).$$

Find its conserved quantity. Simulate trajectories and confirm that they conserve this quantity.

1.8 Write your own code that can calculate a Poincaré section at a predefined (hardcoded) hyperplane corresponding the first variable of a trajectory crossing some constant number (e.g., zero). Use linear interpolation between the trajectory points that "sandwich" the hyperplane to generate the section. Apply your code to, e.g., the Lorenz-63 or Gissinger models. Compare with using the function `poincaresos` from DynamicalSystems.jl and confirm your code works as expected.

1.9 Extend the above code to be able to work with arbitrary hyperplanes of arbitrary orientation. Provide sections for, e.g., the Lorenz-63 system for the planes: $x - y$, $y - z$, $z - x$. *Hint: what is the mathematical equation that defines a hyperplane in a D-dimensional Euclidean space?*

1.10 Create an animation of a Poincaré section that scans a trajectory. Choose a hyperplane of constant orientation, and animate how the section looks as the value that defines the hyperplane changes. For example, if the hyperplane is defined as "when the $x$ coordinate crosses value $v$", then animate the Poincaré section for changing $v$. Apply this animation to the Henon-Héiles system for different trajectories.

1.11 Show that the Fermi-Ulam map

$$u_{n+1} = |u_n + \epsilon \sin\theta_n|, \quad \theta_{n+1} = \left(\theta_n + \frac{2\pi}{|u_n + \epsilon \sin\theta_n|}\right) \quad \mathrm{mod}\ 2\pi$$

is conservative. Evolve a grid of initial conditions and scatter-plot them with different color to get a picture similar to the state space of the standard map, Fig. 1.1. Repeat this for various $\epsilon$ values. Then, analytically find the fixed points of this system and discuss their stability (*Hint: consider $\theta = 0$ or $\theta = \pi$*). Does what you find make sense with respect to the figures you produced?

1.12 Determine the fixed points of the Lorenz-63 system (1.5). Perform a linear stability analysis of the fixed point $(0, 0, 0)$ of the Lorenz-63 system (1.5) for $\sigma, \rho, \beta > 0$ using the Routh-Hurwitz criterion which states: All roots (i.e., Jacobian eigenvalues) of the characteristic polynomial $\pi(\mu) = \det(J_f - \mu I) = a_0 + a_1\mu + a_2\mu^2 + \cdots + a_D\mu^D$ with $a_i \in \mathbb{R}$ (with $I$ the identity matrix) have negative real parts if and only if all determinants

$$C_1 = a_1; \ C_2 = \begin{vmatrix} a_1 & a_0 \\ a_3 & a_2 \end{vmatrix}; \ \dots; \ C_D = \begin{vmatrix} a_1 & a_0 & 0 & 0 & \dots & 0 \\ a_3 & a_2 & a_1 & 0 & \dots & 0 \\ \vdots & \vdots & & & & \\ a_{2D-1} & a_{2D-2} & a_{2D-3} & a_{2D-4} & \dots & a_D \end{vmatrix}$$

(with $a_m = 0$ for $m > D$) have the same sign as $a_d = (-1)^D$. How does the stability of this fixed point depend on the system parameters $\sigma$, $\rho$, and $\beta$?

1.13 Show that a conserved quantity of the system $\dot{x} = xy$, $\dot{y} = -x^2$ is $V = x^2 + y^2$, first by substitution and secondly by separation of variables.

1.14 How can you evolve a continuous system $\dot{\mathbf{x}} = f(\mathbf{x})$ backward in time, i.e., compute the inverse of the flow $\Phi^{-t}(\mathbf{x})$?. Can you do this if $f$ is not invertible? Also solve the same exercise for discrete systems.

1.15 Rewrite the second-order differential equation $\ddot{x} + d\dot{x} - x + x^3 = 0$ of the Duffing oscillator as set of first order differential equations (i.e., for $\sigma.\rho, \beta > 0$ in the standard form of dynamical systems (1.2)). (*Hint: introduce a new state variable $v = \dot{x}$*). Compute the fixed points of this system and determine their stability. Compute and plot the basins of attraction of the stable fixed points for $d = 0.2$ (*Hint: consider a grid of initial values and find out to which attractors the corresponding trajectories will converge*).

1.16 Continuing from above, compute for $d = 0.2$ the stable and unstable manifold of the unstable fixed point and compare the result with the basins. *Hint: evolve a set of initial values located close to the fixed point on the unstable/stable subspace given by the eigenvector corresponding to the unstable/stable eigenvalue. The unstable/stable scenario should be evolved forward/backward in time.*

1.17 The SIR model [11] (also known as Kermack-McKendrick model [12]) describes the evolution of an epidemic

$$\dot{S} = -\beta \frac{SI}{S+I+R}, \ \dot{I} = \beta \frac{SI}{S+I+R} - \gamma I, \ \dot{R} = \gamma I$$

where $S$, $I$, and $R$ are the numbers of susceptible, infected and recovered individuals, respectively. $\beta$ is the infection rate, and $\gamma$ is the rate at which individuals recover. Show that the total number $N = S + I + R$ is a conserved quantity. Implement the model and investigate its dynamics for $\beta = 0.5$ and $\gamma = 0.05$ for initial conditions $(S, I, R) = (999, 1, 0)$ and $(S, I, R) = (990, 10, 0)$ and compare the solutions. Then, simulate a lockdown by reducing the infection rate from $\beta = 0.5$ to $\beta = 0.1$ during a time interval $[10, 30]$.

1.18 Consider the system

$$\dot{v} = -\sin\theta, \ \dot{\theta} = \left(-\cos\theta + v^2\right)/v.$$

Show that it has a conserved quantity and identify it. Evolve and plot a couple of initial conditions and confirm that they follow contour lines of the conserved quantity you found.

1.19 We've mentioned that dissipative systems can have co-existing attractors, just like shown in Fig. 1.3, and just like the exercise with the Duffing oscillator. A very interesting case is the Lorenz-84 system:

$$\dot{x} = -y^2 - z^2 - ax + aF, \quad \dot{y} = xy - y - bxz + G, \quad \dot{z} = bxy + xz - z$$
(1.12)

with $a = 0.25$, $b = 4.0$, $F = 6.886$, $G = 1.337$. For these parameters the state space has an attracting fixed point, an attracting periodic orbit (limit cycle) and a chaotic attractor, all existing at the same time! Find all three of them! *Hint: since you do not have the tools yet to distinguish the different kinds of motion via numerical process, simply plot random initial conditions until you find three where each converges to each of the three attractors.*

# Chapter 2
# Non-chaotic Continuous Dynamics

**Abstract** Continuous systems with only one or two variables cannot be chaotic. However, such systems are used frequently in pretty much all scientific fields as "toy models" that capture the basic relations of certain variables, so it is useful to know how to handle them. Although these non-chaotic systems can be treated analytically to some extend, this is always context-specific. Here we will show how to understand them generally and visually, by looking at their state space. An important tool for performing such a graphical analysis are so-called nullclines, the use of which is demonstrated with an excitable system that exhibits a high amplitude response when perturbed above a relatively small threshold. We also discuss the mental steps one takes while creating a model, and best practices for "preparing" the model equations for study within dynamical systems theory context. Quasiperiodicity is a typical form of non-chaotic motion in higher-dimensional dynamics, and thus serves as a fitting end to this chapter.

## 2.1 Continuous Dynamics in 1D

### 2.1.1 A Simple Model for Earth's Energy Balance

Let's make a simple, 1D continuous system to be used throughout the discussion of this section. This model attempts to provide a dynamical rule for Earth's average temperature $T$. This book is obviously not a climate physics book, so this model will be an oversimplification and should be taken only as a toy example. The precise values of the model parameters will also be somewhat ad hoc (but in general sensible). Nevertheless it is a simple 1D example that ties well with both Chaps. 4 and 12, and provides a small glance at the process of creating a model.

If we want to treat the energy balance of the Earth, we consider two factors: (i) the incoming solar radiation and (ii) the outgoing terrestrial radiation. (i) is equal to $\frac{S_0}{4}(1 - \alpha)$ with $S_0 = 1361$ W/m$^2$ the solar constant, i.e., the direct amount of energy

© The Author(s), under exclusive license to Springer Nature Switzerland AG 2022     21
G. Datseris and U. Parlitz, *Nonlinear Dynamics*, Undergraduate Lecture Notes in Physics,
https://doi.org/10.1007/978-3-030-91032-7_2

that reaches the Earth from the Sun and $\alpha$ the planetary albedo (fraction of solar radiation reflected by the planet as a whole). (ii) is obtained by the Stefan-Boltzmann law, which gives $\epsilon \sigma T^4$ with $\sigma = 5.67 \times 10^{-8}$ W/m$^2$/K$^4$ the Stefan-Boltzmann constant, and $T$ the planet's temperature in Kelvin. $\epsilon$ is the effective emissivity (a fraction between 0 and 1) of the planet, which takes into account the greenhouse effect, which re-uses terrestrial radiation to further heat the Earth.

The balance of these two terms decides whether the planet will further cool or heat, depending on which one is larger. This is expressed in the equation

$$c\frac{dT}{dt} = \frac{S_0}{4}(1 - \alpha(T)) - \epsilon \sigma T^4 \qquad (2.1)$$

where $c$ is the (average) heat capacity of the planet per unit area, in good approximation taken that of a 50 m deep ocean, $c \approx 2 \times 10^8$ J/K/m$^2$. We've explicitly written $\alpha(T)$, assuming that the planetary albedo depends on temperature. On a crude first approximation (and for timescales much longer than a year), a colder planet has more ice, which has a higher albedo. This so-called *ice-albedo feedback* can be expressed with an ad hoc formula $\alpha(T) = 0.5 - 0.2\tanh((T - T_0)/T_r)$, and let's say $T_0 = 263$K, $T_r = 4$K, which means that the albedo varies from 0.7 at temperature of $-20$ degrees °C to 0.3 at 0 °C (Kelvin scale is Celsius scale +273.15). Similar temperature dependence can be assumed for $\epsilon$, but we will not do this here to keep things simple. Equation (2.1) describes a 1D continuous dynamical system.

### 2.1.2 Preparing the Equations

Equation (2.1) has many physical parameters and/or constants, and every quantity involved is also expressed in physical units (e.g., W/m$^2$ or K). While all of this information is extremely important for the context of the equation, it is mostly irrelevant for dynamical systems theory. A good practice for studying a dynamical system is to group and reduce unnecessary parameters, and de-dimensionalize the quantities based on sensible scales defined by the participating parameters and variables. The end product are equations where every symbol is just a real number. The last step, de-dimensionalization, can sometimes be skipped, because the measurement units may already define the most suitable characteristic scale (e.g., in our case 1 K is already the most suitable temperature scale, so there is nothing more to be done). Rescaling time $t \rightarrow \tilde{t} = (S_0/4c)t$ in (2.1) (assuming $S_0$, $c$ to be fixed constants), and renaming $\tilde{t}$ as $t$ results in

$$\frac{dT}{dt} = 0.5 + 0.2\tanh\left(\frac{T - 263}{4}\right) - \epsilon \cdot 10 \cdot (0.002T)^4 \qquad (2.2)$$

where now time is measured in units of $4c/S_0 \approx 0.01$ years and $T$ in Kelvin. Notice also how we wrote $(0.002T)^4$, which leads to better numerical stability due to keeping

**Fig. 2.1** Illustration of the
flow in the state space of the
system (2.2) (indicated by
arrows with size proportional
to the value of $dT/dt$). Fixed
points are denoted by circles:
filled=stable,
empty=unstable

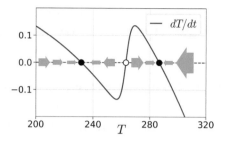

the input argument to values around 1. The only parameter of the model is $\epsilon$ which
we set to 0.65 for now, in a good approximation of current effective emissivity (as
of 2021).

What we've done here also has the advantage of revealing how many *independent*
parameters the system has. In many scenarios it is the ratio of two physical parameters
that matters, not their individual absolute values, although this did not happen here
since we anyways started with a single free parameter $\epsilon$. But most importantly, we
have separated the dynamical context from the physical.

### 2.1.3  Graphical Inspection of 1D Systems

We can immediately determine the long-term dynamics of a 1D system by looking at
its state space. Well, the state space of a 1D continuous systems is just the real line,
so not so interesting, but we can do better! Let's plot $f = dT/dt$ for (2.2) as shown
in Fig. 2.1. If $f > 0$ then $T$ will increase with time, and if $f < 0$, $T$ will decrease.
This is what the arrows in Fig. 2.1 show. Then, it is straightforward to "draw" the
evolution of the system for any initial condition: just follow the arrows.

Specifically to our simplified climate model, we can see from Fig. 2.1 that for
$\epsilon = 0.65$ and starting temperature $0 < T_0 < 260$ the planet goes towards a stable
fixed point representing a cold planet (also called "snowball Earth"), while if $260 <
T_0 < \infty$ we reach the other "warm" equilibrium, which approximately corresponds
to today's climate and albedo, with temperature $T \approx 289$K (as of 2021 at least, it is
likely that in the future things will change).

This graphical approach provides immediate information about how many fixed
points the system has (the zeros of $f$) and if they are stable or not (whether $df/dx$
is negative at the fixed points or not[1]). Finally, a change in parameters means simply
a change of the curve of $f$ in the plot and plotting a new curve will provide the new
fixed points of the system. What becomes clear from this approach is that, no matter

---

[1] In Sect. 1.4 we discussed the criteria for stability based on the eigenvalues of the Jacobian. Here,
since there is only one dimension, the Jacobian is a number, $J \equiv df/dx$, and thus its eigenvalue
coincides with itself.

how complicated function $f$ we "cook up" as the dynamic law, the only thing that can possibly happen in 1D is to arrive at a fixed point or escape to infinity.

## 2.2 Continuous Dynamics in 2D

### 2.2.1 Fixed Points in 2D

In 1D a fixed point can be either repelling or attracting, based on the sign of $df/dx$. For 2D, since there are only two eigenvalues $\mu_1$, $\mu_2$, it is easy to classify the dynamics around fixed points into four categories depending on the values of $\mu_1$, $\mu_2$:

- attractive nodes ($\mu_1$, $\mu_2 < 0$) or repulsive nodes ($\mu_1$, $\mu_2 > 0$)
- attractive spirals ($\mu_1 = \bar{\mu}_2 \in \mathbb{C}$, $\text{Re}(\mu_1) < 0$) or repulsive spirals ($\text{Re}(\mu_1) > 0$) with $\bar{\ }$ the complex conjugate
- hyperbolic (also called saddle) points ($\mu_1 < 0, \mu_2 > 0$)
- centers ($\mu_1 = \bar{\mu}_2 \in \mathbb{C}$, $\text{Re}(\mu_1) = \text{Re}(\mu_2) = 0$)

as shown in Fig. 2.2. Of course, if the state space has only repulsive fixed points and no attractor, then the future of the system is to escape to infinity.

### 2.2.2 Self-sustained Oscillations, Limit Cycles and Phases

Besides fixed points, another structure that can exist in the state space of continuous 2D (or higher dimensional) systems is a *limit cycle*. A limit cycle is an *isolated* periodic orbit (closed curve in state space), as seen in Fig. 2.2 as a black dashed curve. Like the spiral, the limit cycle can also be attractive or repulsive.

Stable limit cycles can occur in autonomous physical systems only if they posses some internal energy source enabling a self-sustained periodic oscillation. Examples are a spring in a mechanical clock or battery supply of an oscillating electronic circuit. Another class of systems that may exhibit period oscillations are chemical reactions far from thermodynamic equilibrium like the *Belousov-Zhabotinsky* (BZ) reaction, where concentrations of reagents vary periodically and can be observed as changing colors. A simplified mathematical model for the BZ reaction is the *Brusselator*

$$\dot{u} = 1 - (b + 1)u + au^2w \tag{2.3}$$

$$\dot{w} = bu - au^2w \tag{2.4}$$

where $u, w \in \mathbb{R}^+$ denote dimensionless concentrations of two of the reactants, the *activator* $u$ and the *inhibitor* $w$, and $a, b \in \mathbb{R}^+$ are parameters representing (normalized) reaction rates. The Brusselator possesses a single fixed point at $(u^*, w^*) = (1, b/a)$. To analyze its stability we compute the Jacobian matrix at

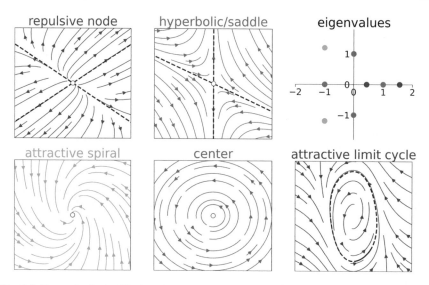

**Fig. 2.2** Dynamics in the 2D plane. The colored arrows are streamplots of $f$ for each case. The panels with purple, orange, cyan, and gray color highlight the possible dynamics around fixed points for 2D continuous dynamical systems. The "eigenvalues" panel, using same colors (there are two dots of each color, since there are two eigenvalues), shows the corresponding Jacobian eigenvalues. The green colored panel displays a stable limit cycle. For all panels the direction of arrows is an example, and counterparts exist with reversed directions (interchanging stable and unstable). For the purple and orange panels, the dashed lines indicate the Jacobian eigenvector directions

$(1, b/a)$

$$J_f(u^*, w^*) = \begin{pmatrix} b - 1 & a \\ -b & -a \end{pmatrix} \tag{2.5}$$

and its trace $\tau = \text{trace}(J_f) = b - a - 1$ and determinant $\delta = \det(J_f) = a$. The real parts of both eigenvalues $\mu_{1,2} = \tau/2 \pm \sqrt{\tau^2/4 - \delta}$ are negative if $\tau = b - a - 1 < 0$ and the fixed point looses its stability at a critical line $b_c = a + 1$ in the $(a, b)$ parameter space. At this line $\tau^2/4 - \delta = -a < 0$ and the eigenvalues thus have non-vanishing imaginary parts which means that the fixed point turns from an attracting spiral into a repulsive spiral. At the same time, this also "gives birth" to a stable limit cycle as illustrated in Fig. 2.3 for $a = 0.3$ and $b = 1.5$. This fundamental transition from an attractive fixed point to an attractive limit cycle is called a *Hopf bifurcation*, and will be discussed in more detail in Chap. 4.

Limit cycles, and periodic orbits in general, are one-dimensional objects in the state space. Therefore, any motion on them can be described by a single variable that increases monotonically from 0 to $2\pi$ and is $2\pi$-periodic, called a *protophase*. A special case of a protophase is the *phase* $\phi$, which increases with a constant rate from $\phi(t_0)$ to $\phi(t_0 + T) = \phi(t_0) + 2\pi$ (mod $2\pi$) during one period $T$ of the oscillation, i.e., $\phi(t) = \phi(t_0) + 2\pi(t - t_0)/(T - t_0)$ (mod $2\pi$). In Fig. 2.3 we show $\phi$ color-coded. The black diamonds on the limit cycle indicate the locations of states

**Fig. 2.3** Limit cycle of the
Brusselator (2.3) occurring
for $a = 0.3$ and $b = 1.5$,
with the phase colorcoded.
The open circle denotes the
location of the unstable fixed
point $(u^*, w^*) = (1, b/a) =
(1, 5)$. At 10 temporally
equidistant states on the limit
cycle we plot black diamonds

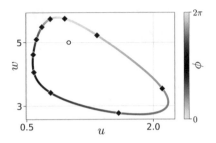

separated by a constant increment in time. As you can see, even though the points
are equidistant in time, they are not equispaced on the limit cycle. That is because
the dynamics at some parts of the cycle can be faster, or slower, than at others. Thus,
the black diamonds are "denser" in parts of the limit cycle where the motion in state
space is slower.

### 2.2.3  Finding a Stable Limit Cycle

The simplest way to find an attractive limit cycle is as follows: simply evolve a bunch
of initial conditions, and see whether at least some of them converge to the *same*
periodic orbit (because a limit cycle must be an isolated periodic orbit). In conjunction
with the techniques that we will introduce in Chap. 3, one can rigorously say whether
the resulting "periodic" orbit is indeed periodic, without even looking at it. However,
one can do slightly better from an analytic perspective. If one can define a connected
region in the state space, say $R$, which does *not* contain any fixed points, but the
vector field ($f$) points "inwards"[2] everywhere at the boundary $\partial R$ of $R$, then $R$ must
contain at least one attractive limit cycle (only valid in 2D of course!).

### 2.2.4  Nullclines and Excitable Systems

In this section we will show how two-dimensional systems can be graphically analyzed be means of so-called nullclines. This approach is particularly useful for studying and understanding excitable systems like the *FitzHugh-Nagumo model*

$$\dot{u} = au(u - b)(1 - u) - w + I$$
$$\dot{w} = \varepsilon(u - w) \tag{2.6}$$

---

[2] Here inwards means that the dot product of $f$ with the normal vector of $R$ at any point $\mathbf{x} \in \partial R$ is
negative.

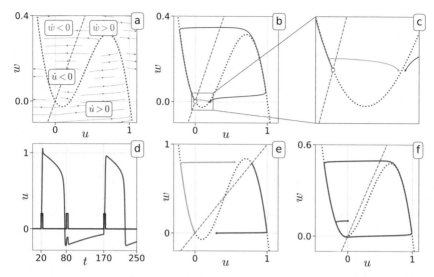

**Fig. 2.4** **a** State space of the FitzHugh-Nagumo model (2.6) for $a = 3, b = 0.2, I = 0$ and $\varepsilon = 0.01$. At the intersection of the nullclines $n_u(u) = au(u - b)(1 - u)$ (green dotted curve) and $n_w(u) = u$ (orange dashed line) a stable fixed point $(0, 0)$ exists. **b** A perturbation pushing the system state to an initial condition at $(0.19, 0)$ results in a trajectory that immediately returns to the fixed point (cyan curve), while a perturbation above threshold $(0.21, 0)$ leads to an excursion before the orbit (purple curve) returns to the origin. **c** Zoom of figure (**b**). **d** Short rectangular pulses $I(t)$ of amplitude 0.2 and width 4 (plotted in black) initiate action potentials of $u$ (i.e., excursions in state space), if the system is excitable. The second pulse, however, does not lead to a high amplitude response, because the system is still refractory. In Code 2.1 we show how we simulated discrete events (pulses, or changes of the parameter $I$) while solving the ODE problem. **e** For $a = 8$, $b = 0.2$, $I = 0$ and $\varepsilon = 0.01$ the nullclines intersect at two points and the system is bistable. **f** With $a = 3$, $b = -0.05$, $I_0 = 0$ and $\varepsilon = 0.01$ the fixed point at the origin becomes unstable and a stable limit cycle occurs (here shown as a purple trajectory starting at $(0, 0.1)$ and converging to the periodic attractor)

which, for example, qualitatively describes the dynamics of neuron firing. In this context, $u$ represents the cell membrane voltage and $w$ is a "recovery" variable, with much slower dynamics (rate of change) than $u$ (we will use $\varepsilon = 0.01$ in the following). $I$ stands for an external injection current and for $I = 0$ the system possesses a fixed point at $(0, 0)$ whose stability can be controlled by the parameters $a$ and $b$. With $a = 3$ and $b = 0.2$ the fixed point $(0, 0)$ is stable and Fig. 2.4a shows for this case a *phase portrait* illustrating typical trajectories. In this figure you can also see the *nullclines*, which are defined as where the individual variables' rate of change becomes zero: $\dot{u} = 0$ (green dotted cubic polynomial) and $\dot{w} = 0$ (orange dashed line).

The nullclines are helpful because they separate the state space into regions where the rates of change of the respective variables are positive or negative, and this allows one to understand the flow of the system. In addition to the nullclines another visualization technique is used in Fig. 2.4 plotting the vector field (2.6) that defines the dynamics with thin black lines and arrows. Most plotting libraries provide a function (typically called "streamlines") to generate such plots.

Why is the FitzHugh-Nagumo model (2.6) called an *excitable system*? To answer this question we have to consider its response to small perturbations of its stable fixed point $(0, 0)$. In general such perturbations could push the state $(u, w)$ in any direction, but in the context of neuro or cardiac dynamics a particular type of perturbation is most interesting: the application of an external current pulse $I(t)$. If this perturbation would consist of a $\delta$-pulse $I(t) = I_0\delta(t - t_0)$ then the state would be shifted[3] $(u, w) \mapsto (u + I_0, w)$ at time $t_0$ and after that the motion would continue as illustrated in Fig. 2.4b, c.

For perturbations below some threshold the resulting trajectory immediately goes back to the fixed point, like the cyan curve. This is what you can typically expect when perturbing a stable fixed point. If, however, the amplitude $I_0$ of the $\delta$-pulse exceeds the threshold value $b$ (given by the zero of the $u$-nullcline $n_u(u) = au(u - b)(1 - u)$) then the trajectory will not immediately return to the fixed point but set out for an excursion in state space, as indicated by the purple curve in Fig. 2.4b, c. This response results in a high peak of the variable $u(t)$ (Fig. 2.4d) which represents in neuroscience or cardiac dynamics a voltage, and in that context the peak is called an *action potential*. During the excursion in state space (purple curve of Fig. 2.4b), as long as the value of the recovery variable $w$ has not yet decayed to zero again, the system is *refractory*, which means that further perturbations have (almost) no impact and only after this refractory period a new excitation is possible (as shown in Fig. 2.4d). This feature leads in extended excitable media to the formation of spiral waves that will be discussed in more detail in Chap. 11.

Actually, we don't need a computer to follow the evolution of the system graphically, provided we have a good impression of the orders of magnitude of the derivatives $\dot{u}$ and $\dot{w}$. Let's look at Fig. 2.4b, c and follow the purple initial condition. Due to the very different time scale set by $\varepsilon \ll 1$ at this part of the state space $\dot{u}$ is much larger than $\dot{w}$ and thus the direction of flow is almost horizontal and to the right (because $\dot{u} > 0$). Then, as we approach the green dotted nullcline more and more, $\dot{u}$ decreases and eventually becomes zero, and thus the trajectory follows a direction which is almost vertical (dominated by $\dot{w}$ and upwards). As the trajectory moves away from the green nullcline, $\dot{u}$ becomes more significant again, and re-orients the flow leftwards. Continuing this process leads us then back to the fixed point.

So far we assumed that the perturbation is given by a $\delta$-pulse applied in the ODE of the first variable. In this idealized case the zero $b$ of nullcline $n_u(u)$ is the threshold for excitation. A similar threshold exists if you use a short (finite) rectangular pulse (or other perturbations), but in this case no simple analytic expression for its value can be derived. An alternative description of the impact of a perturbation is the following. For $I = I_0 > 0$ the point $(0, 0)$ is no longer a fixed point of (2.6) and the state $(u, w)$ will thus start to move during the time interval when $I(t) > 0$ towards the shifted fixed point. Let us assume that at the end of the pulse duration the state is at $(\tilde{u}, \tilde{w})$. Once the pulse is over we are back in the unperturbed system ($I = 0$) and depending

---

[3] Consider $u(t_0 + \Delta t) - u(t_0 - \Delta t) = \int_{t_0-\Delta t}^{t_0+\Delta t} \dot{u}dt = \int_{t_0-\Delta t}^{t_0+\Delta t} au(u - b)(1 - u)dt + \int_{t_0-\Delta t}^{t_0+\Delta t} w\,dt + \int_{t_0-\Delta t}^{t_0+\Delta t} I_0\delta(t - t_0)dt$ which goes to $I_0$ for $\Delta t \to 0$.

on the location of the state $(\tilde{u}, \tilde{w})$ in the diagram shown in Fig. 2.4a the trajectory will immediately return to the fixed point $(0, 0)$ or do an excursion.

Until now the parameters used always led to a single stable fixed point. For $a = 8$, $b = 0.2$, $I = 0$ and $\varepsilon = 0.01$ the nullclines intersect at two points and the system is bistable, i.e., it possesses two stable fixed points (as illustrated in Fig. 2.4e) and perturbations above threshold lead to transitions to the other fixed point. For $a = 3$ and $b = -0.05$ (Fig. 2.4f) the fixed point at $(0, 0)$ looses its stability and a stable limit cycle occurs (due to a Hopf bifurcation, see Sect. 4.1.2 and Further reading).

## 2.3  Poincaré-Bendixon Theorem

We have already seen in Sect. 2.1 that 1D continuous systems can only converge to a fixed point or diverge to infinity. In any case, they can't be periodic or chaotic. 2D continuous systems also cannot have chaotic solutions.

For 2D systems it is not as trivial to show explicitly why there is no chaos. But there is an important theoretical result addressing this topic, the Poincaré-Bendixson theorem. It applies to 2D continuous systems whose state space $S$ is

**Code 2.1** Solving ODEs with events, used to produce Fig. 2.4d.

```
using DynamicalSystems, OrdinaryDiffEq, PyPlot
ds = Systems.fitzhugh_nagumo([0.0,0.0]; I = 0.0)
pulses_start = [20, 80, 170]
pulses_end = pulses_start .+ 4 # 4 = pulse width
pulses = sort!(vcat(pulses_start, pulses_end))
I = 0.2 # strength of pulses of I current

# Create the "callbacks": events in ODE solution
condition(u,t,integ) = t ∈ pulses # trigger condition
function affect!(integ) # what happens at the integrator
    i = integ.t ∈ pulses_start ? I : 0.0
    integ.p[4] = i # 4th parameter is value of current I
end
cb = DiscreteCallback(condition, affect!)

# transform `ds` to form allowing callbacks and solve:
prob = ODEProblem(ds, (0.0, 250.0))
sol = solve(prob, Tsit5(); callback=cb, tstops = pulses)
plot(sol.t, sol[1, :]) # plot timeseries of u
pulse_ts = [any(x -> x ≤ t ≤ x+4, pulses_start) ? I : 0.0
            for t in sol.t]
plot(sol.t, pulse_ts)
```

either the plane, the 2-sphere or the cylinder (see below for 2-torus). It tells us that the state space can be composed only of fixed points and periodic orbits (which limit cycles are). This means that one needs at least three dimensions for chaos in continuous systems. But why is this the case? It is because of the fundamental property of (deterministic) ODE systems, that solutions must be smooth and also must not cross each other in state space (see Sect. 1.1.2). Since they cannot cross themselves, and because the 2D plane is "flat", there is no room for the orbits to wiggle around enough to create chaos (once you read Chap. 3 you will understand that there is not enough room to "fold" in 2D space).

And what about discrete systems you might ask? Here all bets are off! Because there is no smooth continuous line connecting $x_n$ with $x_{n+1}$, there is no limitation coming from the state space dimension itself. Therefore, it is possible to have chaos even with a single dimension in discrete systems! For example, both the logistic map and the standard map that we have presented so far are in fact chaotic systems.

## 2.4 Quasiperiodic Motion

Regular dynamics (i.e., bounded and non-chaotic) are either fixed points, periodic motion, or quasiperiodic motion. But what is this "quasiperiodic" motion? Periodic motion has a frequency $\omega$ and a period $T = 2\pi/\omega$ over which it repeats itself exactly as is forever. What would happen if we combine two frequencies whose ratio is an irrational number? Imagine for a second the following function: $x(t) = \cos(t) + \cos(\sqrt{3}t)$. Is $x(t)$ periodic? No, because there exists no number $T$ such that $x(t + T) = x(t)$. Thus a trajectory containing $x$ as a variable never fully "closes". Is $x(t)$ chaotic? Absolutely not, since it can be written in closed form and doesn't have other properties of deterministic chaos as we will discuss in Chap. 3.

$x(t)$ is an example of *quasiperiodic* motion. $k$th-order quasiperiodicity is the result of the dynamics having $k$ periodic components with rationally independent frequencies.[4] For quasiperiodic motion, each of the system's variables $x_j(t)$ can be expressed as scalar functions of $k$ angle variables (periodic by nature), so that each has its own periodicity. This means that $x_j(t) = h_j(t \mod T_1, t \mod T_2, \ldots, t \mod T_k)$ or equivalently $x_j(t) = h_j(\omega_1 t \mod 2\pi, \ldots, \omega_k t \mod 2\pi) = h_j(\theta_1, \ldots, \theta_k)$ for $\omega_i = 2\pi/T_i$ and $\theta_i = \omega_i t \mod 2\pi$. In our toy example $x(t) = \cos(t) + \cos(\sqrt{3}t)$ we can see that $x(t) = \cos(\theta_1(t)) + \cos(\theta_2(t))$ with $\omega_1 = 1, \omega_2 = \sqrt{3}$. A natural consequence of always being a function of $k$ angles is that the dynamical evolution occurs on a $k$-torus[5] and in the case of dissipative systems the torus can be attractive or repulsive.

To simplify the visualization, let's stick for now with the case of 2nd-order quasiperiodicity, which occurs on a 2-torus. In Fig. 2.5 we show two trajectories,

---

[4] The real numbers $\omega_1, \omega_2, \ldots, \omega_k$ are *rationally independent* if the only $k$-tuple of integers $m_1, m_2, \ldots, m_k \in \mathbb{Z}$ that satisfy the equation $m_1\omega_1 + m_2\omega_2 + \cdots + m_k\omega_k = 0$ is the trivial case of $m_i = 0 \, \forall i$.

[5] A $k$-torus (short for $k$-dimensional torus) is the Cartesian product space of $k$ circles. It is a manifold where each of its dimensions is given by an angle.

**Fig. 2.5** Motion on a torus for two regular trajectories with frequency ratios as shown in the figure. The time evolution is done for exactly $T = 2\pi/\omega_1$, starting from the dashed vertical line. The cyan trajectory with rational frequency ratio (i.e., winding number) is a closed loop while the quasiperiodic trajectory plotted in purple never closes

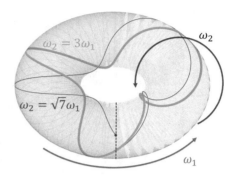

a periodic and a quasiperiodic one. It is useful in parallel to reading this text to also use the live evolution app (online at animations/2/quasiperiodic). As we can see, the periodic version closes perfectly, while the quasiperiodic one does not. In fact, it will slowly fully fill (or cover) the torus if we let the time evolution run for long enough. The frequency ratio $W = \omega_2/\omega_1$ is called the *winding number*, because it counts the average number of rotations in the second direction of the torus the trajectory performs during one revolution along the first direction. It can be used to classify motion into periodic (rational $W$) or quasiperiodic (irrational $W$), as we will see in Chap. 9.

Going back to the general case of $k$-th order quasiperiodicity, an important thing to understand is the power spectrum of $x_j$ (absolute value squared of its Fourier transform). The spectrum of a periodic motion with frequency $\omega$ will, in general, have sharp peaks at all frequencies $m\omega$, $m \in \mathbb{N}$ (but many peaks may be very small and the peak height typically decreases with $|m|$). In the case of $k$th-order quasiperiodicity, there will instead be sharp peaks in every possible linear combination of the $k$ frequencies, $\sum_{i=1}^{k} m_i\omega_i$, $m_i \in \mathbb{Z}$ (but most combinations will have negligible peak height). Keep in mind that when calculating Fourier spectra there will always be two issues to be taken into account: (a) the finite length of any simulated or measured time series leads to a broadening of peaks and (b) measurement noise leads to a background level in the spectrum that could be higher than the peaks you are interested in. Power spectra of chaotic timeseries are fundamentally different, because they are *broadband*: either there are no clear peaks, or if some exist, they are much too broad to be coming from periodic or quasiperiodic motion, and instead indicate predominant frequencies. Example spectra for the three motion types are shown in Fig. 2.6.

Let's return to $x(t) = \cos(t) + \cos(\sqrt{3}t)$ for two reasons. Even though it does capture the essence of quasiperiodicity, it is almost never the case that time evolution of a dynamical can be given in closed form even for quasiperiodic motion. More importantly, we want to point out that only giving $x$ hides the fact that (in general) at least two more dynamic variables $y, z$ exist, both having the same quasiperiodic structure. In the case of continuous dynamics, the Poincaré-Bendixon theorem forbids quasiperiodic motion in typical two-dimensional state spaces. A notable exception to this rule are 2D continuous system where the state space itself is the 2-torus which

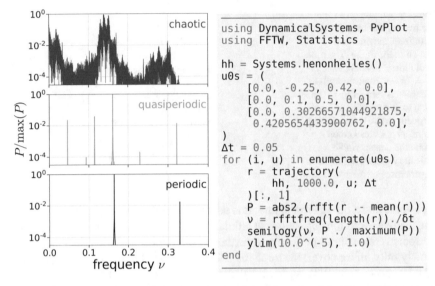

```
using DynamicalSystems, PyPlot
using FFTW, Statistics

hh = Systems.henonheiles()
u0s = (
    [0.0, -0.25, 0.42, 0.0],
    [0.0, 0.1, 0.5, 0.0],
    [0.0, 0.30266571044921875,
        0.4205654433900762, 0.0],
)
Δt = 0.05
for (i, u) in enumerate(u0s)
    r = trajectory(
        hh, 1000.0, u; Δt
    )[:, 1]
    P = abs2.(rfft(r .- mean(r)))
    ν = rfftfreq(length(r))./δt
    semilogy(ν, P ./ maximum(P))
    ylim(10.0^(-5), 1.0)
end
```

**Fig. 2.6** Normalized power spectra for three characteristic trajectories of the Hénon-Heiles system, (1.6). In the spectrum of the quasi-periodic solution 6 peaks are visible which correspond to two fundamental frequencies $\nu_1 = 0.0465$ and $\nu_2 = 0.161$ and the following linear combinations: $2\nu_1 \approx 0.093$, $\nu_2 - \nu_1 \approx 0.115$, $2\nu_2 - 2\nu_1 \approx 0.229$, $2\nu_2 \approx 0.321$. For this case the winding number is simply $W = \nu_1/\nu_2 \approx 3.46$. Notice that the timeseries come from the same trajectories shown in Fig. 1.1

has a topology enabling quasiperiodicity with two incommensurate frequencies (but still no chaos is possible).

Quasiperiodicity exists for discrete systems just as well, without the dimensional limitations of continuous systems. For example, the cyan-colored orbit in the state space of the standard map in Fig. 1.1 is quasiperiodic. See also Sect. 8.1.3.

## Further Reading

The dynamics of non-chaotic continuous systems are analyzed in detail in *Nonlinear dynamics and chaos* by Strogatz [4]. There is also more information about finding limit cycles and distinguishing them from nested periodic orbits as well as connections to index theory. The earliest reference for the Poincaré-Bendixson theorem is [13]. The model (2.1) is inspired by the so-called Budyko-Seller models [14, 15], which express planetary energy balance with a similar type of equations.

The oscillating chemical reaction now known as *Belousov-Zhabotinsky* (BZ) reaction was first observed in the early 1950s by Belousov, but considered to be "impossible" in those days. Therefore, it took him until 1959 to publish the first article on this experiment. Stimulated by this work Zhabotinsky performed additional and more detailed studies which appeared in 1964. Later, in the 1970s Zaikin and Zhabotinsky observed propagation of concentric chemical waves and spiral waves, a topic that we address in Chap. 11. A short summary of the history of the BZ reaction was compiled by Zhabotinsky in Scholarpedia [16] and by another pioneer in the field, Winfree [17]. The Brusselator model (2.3) describing the BZ reaction has been devised by Prigogine and Lefever [18] at the University of Brussels (that's where the name comes from). The notion of protophases has been introduced by Kralemann et al. [19].

The FitzHugh-Nagumo model is a prototypical example of an excitable system. It is named after FitzHugh [20, 21] who devised this system as a simple model for a spiking neuron (solved on an analog computer) and Nagumo [22] who developed an electronic circuit implementation based on a tunnel diode to study propagation of excitations in a chain of 9 such circuits. Later FitzHugh published a computer animation of the propagation of an impulse along a nerve fibre [23]. More details about the history of these developments and discoveries are presented in a Scholarpedia entry [24]. If the perturbations of an excitable system are due to noise the system is now and then kicked above threshold and excursions occur randomly. For noise amplitudes close to the threshold, however, this happens in a relatively regular manner resembling a periodic motion (to some extent). This phenomenon is called *coherence resonance* [25] and provides a mechanism how nonlinearity may increase order in a stochastic process. See also the review article on excitable dynamics by Lindner et al. [26]. More about exitable dynamics in the context of neuro-dynamics can be found in the book *Dynamical Systems in Neuroscience: The Geometry of Excitability and Bursting* by E.M. Izhikevich [27].

Quasiperiodicity, an important concept for both conservative systems, as well as synchronization, is discussed in more depth in the book *Chaos in Dynamical Systems* by Ott [28, Chap. 6].

## Selected Exercises

2.1 For the following 2D systems find and classify the fixed points similarly to Sect. 2.2.1:

$$\dot{x} = 1 + y - e^{-x}, \quad \dot{y} = x^3 - y$$
$$\dot{x} = x^2 y - 3, \quad \dot{y} = y^3 - x$$
$$\dot{x} = \cos(x) + \sin(y), \quad \dot{y} = 2\cos(x) - \sin(y)$$

2.2 Consider the Brusselator (2.3) and find a region $R$ containing the limit cycle such that at its boundary $\partial R$ the vector field $f$ points inwards. This provides a proof that the limit cycle exists in $R$ (see Sect. 2.2.3). *Hint: you may first plot typical trajectories or a streamline plot to get an intuition which shape of $R$ might be appropriate.*

2.3 Analytically calculate the stability of the fixed point of the FitzHugh-Nagumo model as a function of $I$.

2.4 Consider the system $\dot{x} = y$, $\dot{y} = -y(x^2 + y^2 - 1) - x$. Find its fixed points and analyze their stability. Then, analytically show that the system has a limit cycle. Confirm your findings by numerically integrating the system and visualizing the results. *Hint: consider the system in polar coordinates.*

2.5 The Lotka-Volterra model [29] is a well known, but unrealistic,[6] model of predator-prey dynamics given by the equations

$$\dot{x} = \alpha x - \beta xy, \quad \dot{y} = \gamma xy - \delta y$$

where $x$ is the normalized population of the "prey" and $y$ the "predator" and $\alpha, \beta, \gamma, \delta$ positive numbers. First, provide a meaning for each term of the dynamic rule. Then, transform the system into dimensionless form, eliminating as many parameters as possible. Then, draw the nullclines of the resulting system and draw the evolution of a couple of initial conditions without a computer. Do the same visualization on a computer and see whether you were off or not.

2.6 Find a conserved quantity of the Lotka-Volterra model by applying separation of variables. Pick an initial condition, and plot the resulting curve in the state space that corresponds to the conserved quantity. Evolve the initial condition and ensure that it covers the plotted curve.

2.7 Consider an ecosystem with two species that compete for resources but don't have a predator-prey relation. Each species $j$ is characterized by its number $N_j$, its reproduction rate $r_j > 0$, a parameter $w_j < 0$ describing the competition when they interact, and the carrying capacity $K_j > 0$ of the corresponding habitat. The population dynamics is given by $\dot{N}_j = r_j(1 - N_j/K_j)N_j + w_j N_1 N_2$ for $j = 1, 2$. Derive ODEs for the normalized population numbers $n_j = N_j/K_j$ and determine the fixed points of the system and their stability. Which conditions have to be fulfilled such that all fixed points $(n_1^*, n_2^*)$ have $n_j \geq 0$? For which parameter values do you find a stable coexistence of both species and when does one of the species die out (also so-called *competitive exclusion*)? Interpret your results taking into account the meaning of the parameters.

2.8 Simulate the Brusselator model (2.3) and compute the angle $\theta(t)$ of a point $(u(t), w(t))$ with respect to the fixed point $(u^*, w^*) = (1, b/a) = (1, 5)$. Now change the reference point $(1, 5)$ to $(1, 4)$, repeat your computation and report in which way $\theta(t)$ has changed. Also compute $\ell(t)$, which is the arclength the trajectory tracks along the limit cycle (with any point chosen as the start arbitrarily). Normalize the arclength with the length of the limit cycle so that it goes from 0 to $2\pi$. For the Brusselator both $\theta(t)$ and $\ell(t)$ increase monotonically from 0 to $2\pi$ during one period, like $\phi(t)$ of Sect. 2.2.2, and thus are protophases.

---

[6] The Lotka-Volterra is unrealistic because it provides oscillations of arbitrary amplitude depending on initial condition. In reality, while predator and prey populations do follow lagged oscillations with each other's population, there is typically a single characteristic amplitude (limit cycle) that describes the dynamics [30].

However, are they also increasing with a constant speed like the phase $\phi(t)$? *Hint: use an arctangent function with two arguments to compute $\theta$.*

2.9 If only a single variable $u(t)$ from a periodic system is available one use the *Hilbert transform* $v(t) = H[u(t)] = p.v.\frac{1}{\pi}\int_{-\infty}^{\infty}\frac{u(\tau)}{t-\tau}d\tau$ to get a second coordinate $v(t)$ that can be used to compute a protophase. To do so first subtract the mean from your given signal such that $u(t)$ has zero mean. Then compute $v(t)$ and combine $u(t)$ and $v(t)$ to the so-called *analytic signal* $z(t) = u(t) + iv(t)$ in the complex plane where the phase $\psi$ is defined via $z = |z|e^{i\psi}$. Use this approach to compute the phase angle $\psi(t)$ from a $u(t)$ timeseries generated by the Brusselator model (2.3) and compare the result to those obtained with those obtained with other methods to extract a phase. *Hint: you can compute the Hilbert transform by performing a Fourier transform, multiply that by $-i\text{sign}(\omega)$, i.e., shift phases in frequency space by $\pi/2$, and then transform back to the time domain.*

2.10 Simulate the Lorenz-63 system, (1.5) with parameters $\sigma = 10, \beta = 8/3, \rho = 160$, evolving the system first for some transient time and then only keeping the timeseries of the $x$, $y$ variables. The motion in this case is periodic. Use the four ways you have encountered so far to define a (proto)phase: the phase $\phi$, the angle $\theta$ (you decide the origin point), the arclength $\ell$ and the angle $\psi$ of the Hilbert transform of $x$. From these, which ones are actually protophases (i.e., increase *monotonically* with time)? *Hint: it will really help you to plot the orbit in the x-y plane.*

2.11 The Nose-Hoover system [31] is a conservative three dimensional system:

$$\dot{x} = y, \quad \dot{y} = yz - x, \quad \dot{z} = 1 - y^2.$$

In parallel, animate the evolution of the following two initial conditions: $(0, 1.549934227822898, 0)$ and $(0, 1.0, 0)$, the first belonging to a periodic trajectory, and the second resulting in a quasiperiodic trajectory, revolving around the first. Do you see how the quasiperiodic orbit continuously fills a torus? Then, add on the animation one more initial condition: $(0, 0.1, 0)$, which provides chaotic dynamics, and highlights that different parts of the state space can be regular or chaotic.

2.12 Load exercise dataset 5 and for each of its columns plot and analyze the power spectrum. Can you tell which timeseries are periodic, which are quasiperiodic, and which are neither by *only* looking at the power spectra? *Hint: analyze the locations, broadness and clarity of the peaks.*

2.13 Consider the circle map, $x_{n+1} = f(x_n) = (x_n + \alpha) \mod 1$, which maps the interval $[0, 1)$ to itself. Show that for an irrational $\alpha$ every orbit of the map densely fills the interval, i.e., the orbit will come arbitrarily close to every point in the interval. This map produces quasiperiodic trajectories for irrational $\alpha$.

2.14 Consider the Hénon-Heiles system, (1.6) and generate for it three timeseries of the first coordinate using the initial conditions:

$$\{(0.0, -0.31, 0.354, 0.059), (0.0, 0.1, 0.5, 0.0), (0.0, -0.091, 0.46, -0.173)\}.$$

Each timeseries is a 2nd order quasiperiodic motion (two frequencies). Calculate and plot the Fourier spectrum for each timeseries and using the same analysis we did in Fig. 2.6 identify the two fundamental frequencies and calculate the winding numbers.

# Chapter 3
# Defining and Measuring Chaos

**Abstract** Deterministic chaos is difficult to define in precise mathematical terms, but one does not need advanced mathematics to understand its basic ingredients. Here we define chaos through the intuitive concepts of sensitive dependence on initial conditions and stretching and folding. We discuss how volumes grow in the state space using Lyapunov exponents. We then focus on practical ways to compute these exponents and use them to distinguish chaotic from periodic trajectories.

## 3.1 Sensitive Dependence on Initial Conditions

For want of a nail the shoe was lost.
For want of a shoe the horse was lost.
For want of a horse the rider was lost.
For want of a rider the battle was lost.
For want of a battle the kingdom was lost.
And all for the want of a horseshoe nail. [32]

This proverb has been in use over the centuries to illustrate how seemingly minuscule events can have unforeseen and grave consequences. In recent pop culture the term *butterfly effect* is used more often, inspired by the idea that the flap of the wings of a butterfly could lead to the creation of a tornado later on.[1] The butterfly effect is set in the spirit of exaggeration, but as we will see in this chapter, it has a solid scientific basis. It means that tiny differences in the initial condition of a system can lead to large differences as the system evolves. All systems that are chaotic display this effect (and don't worry, we will define chaos concretely in Sect. 3.2.2).

To demonstrate the effect, we consider the Lorenz-63 model from (1.5). In Fig. 3.1 we create three initial conditions with small differences between them and evolve all three of them forward in time, plotting them in purple, orange and

---

[1] A flap of butterfly wings can never have such consequences. The statement is more poetic than scientific.

© The Author(s), under exclusive license to Springer Nature Switzerland AG 2022
G. Datseris and U. Parlitz, *Nonlinear Dynamics*, Undergraduate Lecture Notes in Physics,
https://doi.org/10.1007/978-3-030-91032-7_3

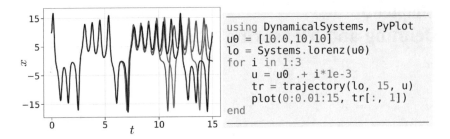

```
using DynamicalSystems, PyPlot
u0 = [10.0,10,10]
lo = Systems.lorenz(u0)
for i in 1:3
    u = u0 .+ i*1e-3
    tr = trajectory(lo, 15, u)
    plot(0:0.01:15, tr[:, 1])
end
```

**Fig. 3.1** Sensitive dependence on the initial condition illustrated for the Lorenz-63 system simulated with three different, but very similar, initial conditions

black color in Fig. 3.1. Initially there is no visual difference, and only the black curve (plotted last) is seen. But after some time $t \approx 8$ the three different trajectories become completely separated. A 3D animation of this is available online at animations/3/trajectory_divergence. It is also important to note that decreasing the initial distance between the trajectories would not eliminate the effect. It would take longer, but eventually the trajectories would still separate (see Sect. 3.1.2). This is the simplest demonstration of *sensitive dependence on the initial condition*, the mathematically precise term to describe the butterfly effect. It means that the *exact* evolution of a dynamical system depends sensitively on the *exact* initial condition. It should be noted however that if we are interested in averaged properties of a dynamical system instead of the exact evolution of a single trajectory starting at some particular initial value, then sensitive dependence isn't much of a problem due to ergodicity, see Chap. 8.

### 3.1.1  Largest Lyapunov Exponent

Let's imagine a state $\mathbf{x}(t)$ of a chaotic dynamical system evolving in time. At time $t = 0$ we create another trajectory by perturbing the original one with a small perturbation $\mathbf{y}$, with $\delta = ||\mathbf{y}||$, as shown in Fig. 3.2. If the dynamics is chaotic, the original and the perturbed trajectories separate more and more in time as they evolve. For small perturbations, this separation is happening approximately exponentially fast. After some time $t$ the two trajectories will have a distance $\delta(t) \approx \delta_0 \exp(\lambda_1 t)$. The quantity $\lambda_1$ is called the *largest Lyapunov exponent* and quantifies the *exponential divergence* of nearby trajectories.

After some time of exponential divergence, the distance between the two trajectories will saturate and stop increasing exponentially, as is also visible in Fig. 3.2. Precisely why this happens will be explained in more detail in Sect. 3.2.4, but it stems from the fact that $f$ is nonlinear and the asymptotic dynamics is bounded in state space. The important point here is that *exponential* divergence of nearby trajectories occurs *only for small distances between trajectories*. This fact is the basis of

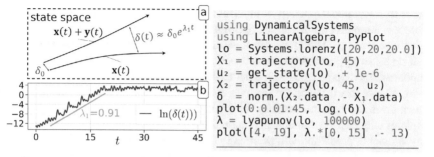

**Fig. 3.2** **a** Sketch of exponential divergence of neighbouring trajectories characterized by the largest Lyapunov exponent $\lambda_1$. **b** The evolution of $\delta(t)$ for the Lorenz-63 system, produced via the code on the right. In general, the initial condition should be on or very near to the set whose divergence properties are to be characterized. For the Lorenz-63 model this requirement is not so strict because it is strongly dissipative such that any trajectory converges very quickly to the chaotic attractor

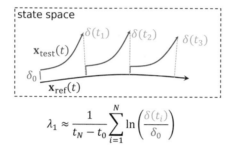

$$\lambda_1 \approx \frac{1}{t_N - t_0} \sum_{i=1}^{N} \ln\left(\frac{\delta(t_i)}{\delta_0}\right)$$

**Fig. 3.3** Computing the largest Lyapunov exponent $\lambda_1$. The initial distance $\delta_0$ of the two trajectories $\mathbf{x}_{\text{ref}}$ and $\mathbf{x}_{\text{test}}$ which are evolved in parallel has to be chosen very small. Their distance increases with time, and at times $t_i = i \cdot \Delta t$ the orbit $\mathbf{x}_{\text{test}}$ is re-scaled to have again distance $\delta_0$ from $\mathbf{x}_{\text{ref}}$ with the transformation $\mathbf{x}_{\text{test}} \rightarrow \mathbf{x}_{\text{ref}} + (\mathbf{x}_{\text{test}} - \mathbf{x}_{\text{ref}})(\delta_0/\delta(t_i))$. $\lambda_1$ is approximated by time-averaging the logarithm of the ratio of distances

the algorithm described in Fig. 3.3, which estimates $\lambda_1$, and is also implemented as `lyapunov` in DynamicalSystems.jl.

## 3.1.2 Predictability Horizon

The value of $\lambda_1$ has immediate practical impact. Imagine for a moment that there exists a "real" physical system that we want to predict its future. Let's also assume that we have a mathematical model that perfectly describes the real system. This is never true in reality, and the imperfect mathematical model introduces further uncertainty, but for now let's just pretend.

We now make a measurement $\tilde{\mathbf{x}}$ of the state of the real system $\mathbf{x}$. Our goal is to, e.g., plug $\tilde{\mathbf{x}}$ into our mathematical model and evolve it in time on a computer, much faster than it would happen in reality. But there is a problem: our measurement of the state $\mathbf{x}$ is not perfect. For various reasons the measurement $\tilde{\mathbf{x}}$ is accompanied by an error $\mathbf{y}$, $\tilde{\mathbf{x}} = \mathbf{x} + \mathbf{y}$. Therefore, our measured state $\mathbf{y}$ has an initial distance from the real state $\delta = ||\mathbf{y}||$. If the system is chaotic, the measured state and the real state will exponentially diverge from each other as they evolve in time as we have discussed so far. Let's say that if the orbits separate by some tolerance $\Delta$, our prediction has too much error and becomes useless. This will happen after time $t_\Delta = \ln(\Delta/\delta)/\lambda_1$. It explains why chaotic systems are "unpredictable", even though deterministic, since any attempt at prediction will eventually become unusable.

The characteristic time scale $1/\lambda_1$ is often called the *Lyapunov time* and defines the *predictability horizon* $t_\Delta$. One more difficulty for predicting chaotic systems is that because of the definition of $t_\Delta$, in order to increase this horizon linearly (e.g., double it), you need to increase the measurement accuracy exponentially (e.g., square it). Although the reasoning regarding the predictability horizon is valid so far, we should be careful with how generally we apply it, see Chap. 12 for further discussion.

## 3.2  Fate of State Space Volumes

### 3.2.1  Evolution of an Infinitesimal Uncertainty Volume

So far we have represented "uncertainty" of a measurement as a single perturbation of an initial condition. While this is useful for conceptualizing divergence of trajectories, a more realistic scenario is to represent uncertainty as an infinitesimal volume of perturbed initial conditions around the state of interest $\mathbf{x}_0$. Then, we are interested to see how this infinitesimal volume changes size and shape as we evolve the system.

To simplify, we will consider the initial volume as a $D$-dimensional hypersphere around an initial condition $\mathbf{x}_0$. Once we evolve the infinitesimal hypersphere, we will see it being deformed into an ellipsoid as shown in Fig. 3.4. Movies of this process for continuous and discrete systems in 3D can be seen online in animations/3/volume_growth. The ellipsoid will also rotate over time, but for this discussion we only care about the change in its size. We can quantify the change

**Fig. 3.4** Evolution of an infinitesimal perturbation disk along a trajectory of a two-dimensional discrete system. From step 1 to 2, the purple vector gets mapped to the orange one (because arbitrary perturbations align towards the maximal expansion direction)

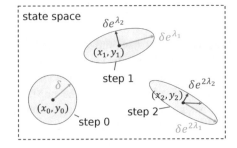

in size by looking at the growth, or shrinking, of the axes of the ellipsoid. What we find, for almost all initial conditions $x_0$, is that the $i$-th axis of the ellipsoid will approximately grow, or shrink, exponentially fast in time, as $\delta e^{\lambda_i t}$ with $\delta$ being the initial radius of the infinitesimal hypersphere. Keep in mind that this will only occur for small perturbation volumes. Evolving a hypersphere with a finite-sized, but still small, radius $\delta$ means that the exponential increase along the directions with $\lambda_i > 0$ will sooner or later saturate, exactly as it has been discussed in the Sect. 3.1.1.

### 3.2.2 Lyapunov Spectrum

The exponential rates $\lambda_i$ along each of the $D$ dimensions of this hypersphere are called the *Lyapunov exponents* or the *Lyapunov spectrum*, typically ordered in decreasing order[2] $\lambda_1 \geq \lambda_2 \geq \cdots \geq \lambda_D$. The largest exponent $\lambda_1 \equiv \lambda$, which we already discussed in the previous sections, is in some sense the most important because over time it will trump all other rates of change. Nevertheless the entire Lyapunov spectrum is important, and quantifies several dynamic characteristics (see below). The spectrum can also be used to give a quantitative definition of deterministic chaos: if the dynamic evolution yields at least one positive exponent Lyapunov exponent, and the system state stays bounded in state space, then this means deterministic chaos.

In Sect. 1.5 we discussed how the Jacobian of the dynamic rule $f$ provides the rate of change (for discrete systems the direct change) of infinitesimal perturbations around a specific point in the state space $x$. Since Lyapunov exponents quantify stretching (or contracting) of perturbations in characteristic directions, we can in principle obtain all exponents if we can find out how these perturbations evolve. In this section we will provide a high-level, and somewhat ad hoc description of an algorithm to calculate the exponents. We recommend however to have a look at Appendix A, where we present the algorithm more explicitly, exposing some fundamental details that lead to (3.2).

To calculate the $k$ largest Lyapunov exponents we represent a $k$-dimensional volume around a point $x$ with $k \leq D$ perturbation (also called deviation) vectors, equivalently formulated as a $(D, k)$ matrix $Y$. Think of these vectors as the axes of a $k$-dimensional parallelepiped. For each perturbation vector $y$ (i.e., each column) we know that its instantaneous rate of change is given by $\dot{y} = J_f(x) \cdot y$, or $y_{n+1} = J_f(x) \cdot y_n$ for discrete systems, as was shown in Sect. 1.4. Thus, the full evolution of $Y$ can be written as

$$\begin{aligned} \dot{x} &= f(x) \\ \dot{Y} &= J_f(x) \cdot Y \end{aligned} \quad \text{or} \quad \begin{aligned} x_{n+1} &= f(x_n) \\ Y_{n+1} &= J_f(x_n) \cdot Y_n. \end{aligned} \tag{3.1}$$

---

[2] As the relation $\geq$ indicates, the Lyapunov spectrum can be (partly) degenerated (with some $\lambda_{i1} = \lambda_{i2} = \ldots = \lambda_{ij}$). To simplify illustrations and discussion, however, we will in the following mainly focus on the case $\lambda_1 > \lambda_2 > \cdots > \lambda_D$, but all results are generalizable and true also for degenerated spectra.

Starting at $\mathbf{x}_0$ with initial perturbations $Y_0$ being an arbitrary orthonormal matrix, we can evolve $Y(t)$ in parallel with the state $\mathbf{x}$ so that the state-dependent Jacobian $J_f(\mathbf{x}(t))$ is used at each point along the trajectory. Equations (3.1) are called the equations for the linearized dynamics or for the dynamics in tangent space.

Alright, now you might think that we're pretty much done and computing all Lyapunov exponents is straightforward. You solve simultaneously (3.1) for the dynamical system of interest and obtain how the initial volume $Y_0$ changes in time to $Y(t)$. Then you examine $Y(t)$ and see how its volume has increased or decreased over time along characteristic dimensions. A suitable way to do this is using the QR-decomposition method, $Y(t) = Q(t)R(t)$, which decomposes $Y$ into an column-orthogonal $(D, k)$ matrix $Q$ and an upper-triangular $(k, k)$ matrix $R$ with $R_{ii} > 0$. Think of $Q$ as representing how $Y$ has rotated over time, while $R$ (and especially the diagonal entries of $R$) indicates how $Y$ has changed size over time along each of the $k$ dimensions. Given that the Lyapunov exponents are quantifying the logarithmic increase of size of $Y$, we can express them as (see Appendix A)

$$\lambda_i = \lim_{t \to \infty} \frac{1}{t} \ln R_{ii}(t). \tag{3.2}$$

Provided that we have evolved $Y(t)$ for long enough (to sufficiently sample all of the chaotic set), we simply decompose it to obtain $R(t)$, apply (3.2), and we're done calculating the Lyapunov spectrum.

This is in principle correct, but it will fail in practice. The problem is that you cannot compute $Y(t)$ for large $t$ for two reasons. First, the elements of this matrix will increase or decrease exponentially in time and you will quickly run into problems with over- or underflow (due to the finite representation of numbers in a computer). Second, almost all initial perturbations will converge towards the most quickly expanding direction (see orange vector in Fig. 3.4). Although you initialize (3.1) with an orthonormal matrix $Y_0$, the column vectors of $Y(t)$ will not remain orthogonal during their evolution but converge to the direction corresponding to $\lambda_1$. Thus, after some time your solution for $Y(t)$ is effectively useless, because all its column vectors have become almost colinear.

Fortunately, it is easy to avoid these issues by remembering that the linearized dynamics does not care about the actual size of $Y$, only its shape. Thus we can split the long simulation interval $[0, T]$ into shorter $\Delta t$-sized intervals and perform some renormalization and reorthogonalization of solution vectors after each time interval. Conveniently, we can use $Q(t)$, since it is already orthonormal, and replace $Y$ with $Q$ after every $\Delta t$ step. We only need to make sure that we accumulate somewhere the values of $\log(R_{ii})$, since the exponents after $N$ steps will be approximated by $\lambda_i = \frac{1}{N\Delta t} \sum_i \log(R_{ii}(t = i\Delta t))$. A Julia implementation of this algorithm is presented in Code 3.1. The object `tangent_integrator` solves (3.1) iteratively while at each step the $Q$ matrix is set as the $Y$ initial matrix and the diagonal entries of $R$ are accumulated.

**Code 3.1** Computation of the `k` largest Lyapunov exponents of the dynamical system `ds`, by re-scaling $k$ deviation vectors every `Δt` units of time for a total of `N` times.

```
using DynamicalSystems, LinearAlgebra
function lyapunov_exponents(ds::DynamicalSystem, N, k, Δt)
    integ = tangent_integrator(ds, k)
    t0, λ = integ.t, zeros(k)
    for i in 2:N
        step!(integ, Δt)
        Y = get_deviations(integ)
        qrdec = qr(Y)
        for j in 1:k
            λ[j] += log(abs(qrdec.R[j,j]))
        end
        # Set current Q as new Y
        set_deviations!(integ, qrdec.Q)
    end
    λ ./= (integ.t - t0)
    return λ
end
```

If one does not care about finding rates of growth across different dimensions, but only about the time evolution of the volume $v$ of an infinitesimal ball around some $\mathbf{x}_0$ as a trajectory from $\mathbf{x}_0$ evolves, one can use the equations we have already seen in Sect. 1.5. Specifically, it holds that $\dot{v} = \text{trace}\left(J_f(\mathbf{x}(t)) \cdot v\right)$ for continuous systems and $v_{n+1} = \det\left(J_f(\mathbf{x}_n)\right) \cdot v_n$ for discrete (see Appendix A for a proof). As above, these equations need to be integrated in parallel with the dynamic rule to correctly obtain $\mathbf{x}$ as a function of time.

### 3.2.3 Properties of the Lyapunov Exponents

Before we discuss formal properties of the exponents, we need to stress again that they arise from the *linearized* dynamics. Therefore, they quantify perturbation/error growth that is small enough that it can be adequately described by the linearized dynamics. In high-dimensional multi-component systems, like the weather, it can happen that a typical error is already beyond this linear regime and thus these Lyapunov exponents (as we defined them here) are of limited use for providing predictability limits. This we will discuss more in Chap. 12.

Now onto formal properties. For any invariant set of a dynamical system a unique Lyapunov spectrum $\{\lambda_1, \ldots, \lambda_D\}$ exists. Almost all initial states $\mathbf{x}_0$ from this set provide the same values for the Lyapunov exponents in the limit of infinite time in (3.2). Arguably the most important property of the Lyapunov exponents $\{\lambda_i\}$ is that they are true characteristics of the set because they are dynamic invariants, e.g., transforming the system via diffeomorphisms preserves the exponents. Such dynamic invariance does not hold for, e.g., statistical moments like means or standard deviations. Another notable property is that for continuous systems at least one exponent $\lambda_i$ always equals zero.[3] This zero exponent is the one associated with the direction "along the flow", as perturbations along the direction of the trajectory never decrease or increase (on average). For the case of $k$-order quasiperiodicity (Sect. 2.4) $k$ exponents are equal to zero (and this is true irrespectively if the system is discrete or continuous). If $k < D$, the remaining exponents are either $<0$ (dissipative systems) or 0 (conservative systems).

The sum of all $D$ Lyapunov exponents is equal to the (average) volume growth rate in the state space. Thus, for dissipative systems this sum is negative, while for conservative systems the sum of Lyapunov exponents is always 0. In fact, specifically for Hamiltonian mechanics it holds $\lambda_i = -\lambda_{D-i+1}$, due to the conjugate symmetry of Hamilton's equations of motion. Last but not least, Lyapunov exponents provide a practical way to estimate the dimension of high dimensional attractors, which we discuss further in Sect. 5.4.4.

### 3.2.4  Essence of Chaos: Stretching and Folding

The concept of the sensitive dependence on initial conditions, stemming from the presence of at least one positive Lyapunov exponent, is a characteristic of all chaotic systems, irrespective of how many dimensions they have or if they are discrete or continuous. But this sensitive dependence is not sufficient for a definition of chaos. In fact, linear systems can also display sensitive dependence, when their defining matrix has at least one unstable eigenvalue (see Sect. 1.4), but then of course all trajectories will simply diverge to infinity and not stay bounded in state space.

What makes the chaotic systems "chaotic" is that when evolving in time they all in some way perform the following action: the state space is first stretched and then folded within itself. *This* is the defining characteristic of chaos, and we sketch the process in Fig. 3.5. The stretching part is where the sensitive dependence comes from, and it occurs due to the locally linearized dynamics being "unstable", in the sense of having at least one expanding direction, i.e., positive Lyapunov exponent. However, as state space volumes increase in size, the linearized dynamics no longer hold. Then the nonlinear dynamics takes over and *folds* the state space back onto itself, keeping the dynamics bounded in state space. That is why only nonlinear systems can be chaotic; linear systems with positive Lyapunov exponents always escape to infinity.

To illustrate stretching and folding in commonly used dynamical systems like those of Sect. 1.1.1 is not so easy. One can intuitively use billiards (Chap. 8), and we

---

[3] An exception here are fixed points whose exponents are directly derived from the Jacobian eigenvalues at the fixed point, see exercises.

**Fig. 3.5** A sketch illustrating the process of stretching and folding. An initial set (grey color) is stretched and folded upon itself in every iteration. In principle this figure is the same process as kneading dough, which the dedicated lecturer can demonstrate in class: add two concentrated colorant droplets in the dough and start kneading to illustrate chaos!

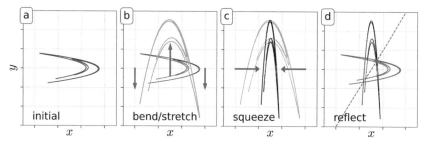

**Fig. 3.6** Stretching and folding in the Hénon map. An animation of this process can be found online at animations/3/henon_folding

made a video for this available online.[4] Thankfully we can show how stretching and folding happens in another commonly used discrete system, the Hénon map

$$x_{n+1} = 1 - ax_n^2 + by_n, \quad y_{n+1} = x_n \tag{3.3}$$

(we use $a = 1.4$ and $b = 0.3$ unless noted otherwise). It is a chaotic system with a chaotic attractor shown in Fig. 3.6a, while being discrete and thus easy to simulate (see Sect. 5.3.2 for more on chaotic attractors). The generating equations for this map can be considered step by step as follows. First, the $y$ coordinate is stretched and folded at the same time, because it is transformed as $(x, y) \mapsto (x, 1 - ax^2 + y)$ (Fig. 3.6b). Then the $x$ coordinate is squeezed as $(x, y) \mapsto (bx, y)$ (Fig. 3.6c). A final reflection $(x, y) \mapsto (y, x)$ shows how the attractor is constructed (Fig. 3.6d).

### 3.2.5 Distinguishing Chaotic and Regular Evolution

The simplest practical application of what has been discussed so far is the predictability horizon, Sect. 3.1.2. Another practical application is determining whether a given trajectory or timeseries represents chaotic time evolution or not. This is useful

---

[4] https://www.youtube.com/watch?v=svV1MsUdInE.

**Fig. 3.7** Chaoticity map for
the Hénon-Heiles system,
(1.6), using the maximum
Lyapunov exponent of each
initial condition to color it,
and the Poincaré surface of
section for visualization.
Initial conditions with dark
purple color are regular,
while the rest are chaotic

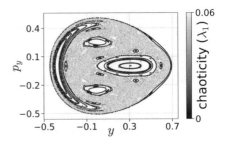

in several occasions. First, in many dynamical systems (especially in those arising
from Hamiltonian mechanics) different initial conditions result in chaotic or regular
motion, and one wants to distinguish both. Second, if we are given an unknown
(measured) timeseries or trajectory, we cannot know a priori whether it represents
regular or chaotic motion, and have to analyze further. Distinguishing chaotic from
stochastic timeseries is a much harder task that we discuss in Chap. 7.

As mentioned in Sect. 3.1.1, $\lambda_1$ quantifies whether the motion is chaotic ($\lambda_1 > 0$)
or regular ($\lambda_1 \leq 0$), assuming of course always that the time evolution stays within
some finite bounds for all times. We can use this fact to create "chaoticity" maps for
the state space of a dynamical system. We start by evolving several initial conditions
and for each one we color code it according to its $\lambda_1$. Then a suitable visualization
is done. In Fig. 3.7 we did this for the Hénon-Heiles system, (1.6), and we used the
Poincaré surface of section (Sect. 1.6) for visualization. An interactive app of this
plot is available online at animations/3/poincare_lyapunov.

## 3.3  Localizing Initial Conditions Using Chaos

Chaotic dynamics limits our ability to forecast temporal evolution of a system as
discussed in Sect. 3.1.2. But conversely, sensitive dependence on the initial values also generates information! More precisely, information about the location of
initial states, if the future evolution of the system is known. A simple example
to illustrate this feature is the *Bernoulli map* $x_{n+1} = 2x_n$ (mod 1) which generates chaotic orbits with Lyapunov exponent $\lambda = \log(2)$. If you use a binary representation $x = \sum_{j=1}^{\infty} b_j 2^{-j} \equiv 0.b_1 b_2 b_3...$ of the state variable $x \in (0, 1)$ the corresponding sequence of zeros and ones is shifted to the left (due to the multiplica-
tion by 2) and the leading digit is removed by the modulo operation (for example:
$0.1100101 \rightarrow 0.100101$). So if you record or observe only the first $k$ digits of the
current state $x_n$ of this chaotic system you will learn about the value of the $k + 1$-th
digit of $x_n$ when measuring $x_{n+1}$ (after the next iteration), about the $k + 2$-th digit of
$x_n$ after the second iteration and so forth. The iterations shift digits corresponding to
tiny variations $2^{-m}$ ($m \gg k$) of the initial state to the left so that they finally become

**Fig. 3.8** Two states $\mathbf{x}_0$, $\mathbf{y}_0$
are evolved and sampled at
times $t = 0, \Delta t, 2\Delta t$ (open
circles). While $\mathbf{x}_0$ resides in
the intersection $B_0 \cup$
$\Phi^{-\Delta t}(B_1) \cup \Phi^{-2\Delta t}(B_2)$, $\mathbf{y}_0$
is not in this subset, but only
in $B_0 \cup \Phi^{-\Delta t}(B_1)$. This
allows us to distinguish the
two with more precision than
our initial resolution $\varepsilon$

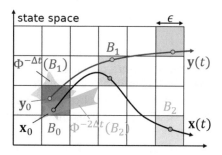

"visible" with your observational resolution of $k$ digits. Thus, following the orbit
you learn more and more about the precise location of the initial condition.

**Code 3.2** Code used to generate Fig. 3.7

```
using DynamicalSystems, PyPlot
hh = Systems.henonheiles()
# create some initial conditions, all at energy = 0.13
ics = Systems.henonheiles_ics(0.13, 15)
cmap = matplotlib.cm.get_cmap("viridis")
for ic in ics
    psos = poincaresos(hh, (1, 0.0), 2000; u0 = ic)
    λ = lyapunov(hh, 10000; u0 = ic)
    v = clamp(λ/0.06, 0, 1)
    scatter(psos[:, 2], psos[:, 4]; color = cmap(v))
end
```

This mechanism works not only with this special type of a 1D chaotic map but for
any chaotic system, as illustrated in Fig. 3.8. Let's assume that you can measure states
of your system of interest only with resolution $\varepsilon$, i.e., you can only tell in which box of
an $\varepsilon$-partition of the state space this state is located. Another different state in the same
box can thus *not* be distinguished from the first state. Consider an orbit starting at
some state $\mathbf{x}_0$ which is sampled at times $n\Delta t$ resulting in a sequence of states $\mathbf{x}_0$, $\mathbf{x}_1 =$
$\Phi^{\Delta t}(\mathbf{x}_0), \ldots, \mathbf{x}_n = \Phi^{n\Delta t}(\mathbf{x}_0)$ visiting different boxes $B_0, B_1, \ldots, B_n$. Following the
inverse flow $\Phi^{-i\Delta t}$ we know that the initial state $\mathbf{x}_0$ has to be located in the intersection
$B_0 \cup \Phi^{-\Delta t}(B_1) \cup \ldots \cup \Phi^{-n\Delta t}(B_n)$ whose volume shrinks for increasing $n$ in case of
a chaotic system. Thus, the longer the trajectory, the smaller the intersection and the
more information we obtain about the location of $\mathbf{x}_0$, which allows us to distinguish
it from other initial conditions $\mathbf{y}_0$ that start in the same box.

Crucial for this distinction is that the dynamics is chaotic and thus trajectories
diverge during time evolution. This process of information generation is quantified
by the *Kolmogorov-Sinai (KS) entropy*. However, its formal definition exceeds the
scope of the book, and we haven't discussed the concept of entropy yet either. See
Further reading of Chap. 5 for more on this.

## Further Reading

Poincaré is without a doubt one of the most important figures for the theory of dynamical systems, nonlinear dynamics and deterministic chaos. Not only he had a plethora of contributions of tremendous impact, including some of the most foundational theorems, but also he was also the first to discover sensitive dependence on initial conditions and deterministic chaos [33] (some sources wrongly attribute this to Lorenz, see below). He first encountered sensitive dependence in the context of the three body problem. Poincaré's goal was to prove that the solar system (represented by three gravitationally attractive bodies, Sun, Moon and Earth) was stable. To progress with his proof, at some point he made the assumption that trajectories that were initially close, would stay close indefinitely. This seemed reasonable at first, given that until that time only regular dynamics had been studied. Thankfully for us, Poincaré realised that this assumption was false, see the article by Green [34] for the interesting story behind these events. What Poincaré understood was that the three body system is chaotic and thus displays sensitive dependence on initial conditions.

The term "butterfly effect" is originating from a title of a talk of Lorenz, "Does the flap of a butterfly's wings in Brazil set off a tornado in Texas?" which is based on his monumental paper on deterministic nonperiodic flow [8]. Lorenz, while not being the first to discover deterministic chaos, is often named the "father" of chaos theory. This is justified, because Lorenz's discoveries and discussions made deterministic chaos be understood as the widespread and almost universal phenomenon that it is, thus establishing the importance of nonlinear dynamics and chaos in the scientific community. Lorenz was also the first person to rigorously discuss and appreciate the impact of chaos and sensitive dependence on initial conditions for the real world, with extra focus on what this means for weather and climate. For more about E. Lorenz, see the discussion by Ott [35].

A relevant textbook that is written in a mathematical tone with proofs and deeper information related to deterministic chaos and its definition is *Chaos: An introduction to Dynamical Systems* by Alligood, Sauer and Yorke. See also *Dynamics: the geometry of behavior* by Abraham and Shaw [36] for several illustrative demonstrations of chaos and stretching and folding.

Lyapunov had a lot of contributions and involvement in the study of stability of dynamical systems [37], hence the Lyapunov exponents bear his name. See also the relevant Scholarpedia page [38] for more historical references. The existence and other properties of Lyapunov exponents are rigorous mathematical results, coming from the Oseledets theorem [39]. The practical method of calculating Lyapunov exponents that we presented here is based on the work of Shimada and Nagashima [40] and Benettin et al. [41]. See [42] for an illustrative description of the algorithm, [43] for further details and alternative algorithms, and [44] for a more recent review on Lyapunov exponents. The most comprehensive treatment of Lyapunov exponents and vectors can be found in book written by Pikovsky and Politi devoted only to this topic [45].

In this chapter we mentioned that at least one positive Lyapunov exponent and bounded evolution of a nonlinear dynamical system means deterministic chaos. Precisely defining deterministic chaos is unfortunately a bit more complicated than that. A nice discussion of the shortcomings of the various previous definitions of chaos is given by Hunt and Ott in [46]. The authors there define an improved definition of chaos based on a quantity called expansion entropy, which is also available in DynamicalSystems.jl as `expansionentropy`.

Besides the Lyapunov exponents there are several other methods that can quantify chaos based on the linearized (or tangent) dynamics. One of them is GALI [47], which is implemented as `gali` in DynamicalSystems.jl. GALI looks at how fast the various perturbation vectors that compose volumes in the linearized space align towards the direction of maximal increase. This alignment will happen exponentially fast for chaotic motion, or polynomially fast (or no alignment at all) for regular motion. Other methods that can be used as chaotic quantifiers are the Orthogonal Fast Lyapunov Indicators (OFLI) [48, 49] or the Mean Exponential Growth of Nearby Orbits (MEGNO) [50]. These methods (and others) are discussed in detail in *Chaos Detection and Predictability* [51] (several authors). Articles that compare the efficiency of methods to quantify chaos are also cited in this book. For example, a comparative study of all methods that look at perturbation growth in the linearized space is [52]. Worth mentioning is a recent different quantitative description of chaos still requiring a known dynamic rule by Wernecke et al. [53]. Instead of using the linearized dynamics here one evolves several nearby trajectories in the real state space and looks at their cross-correlations. It is intuitive (and also can be shown explicitly) that correlation functions of chaotic systems decay exponentially simply because of the exponential separation of nearby orbits. Based on this, this method can claim not only if the trajectory is chaotic or not, but also to what degree it is chaotic (by looking at how the correlations decay). This method is implemented in DynamicalSystems.jl as `predictability`.

## Selected Exercises

3.1 Write your own function that takes as an input a discrete dynamical system and performs the algorithm described by Fig. 3.3 to calculate the maximum Lyapunov exponent $\lambda_1$. Apply it to the standard map (1.4) and calculate its $\lambda_1$ for $k$ ranging from 0.6 to 1.2, always starting with initial condition (0.1, 0.11). *Hint: the chosen test state does not really matter, provided that it is chosen close enough to the reference state.*

3.2 Repeat the above exercise, but now for the Lorenz-63 system for $\rho$ ranging from 20 to 100. Since this is a continuous system, and you need to solve ODEs, you can do it as follows: evolve the two initial conditions for $\Delta t$, then stop integration, obtain the final states, and then re-normalize the test one as in Fig. 3.3. Then set this as a new initial condition and evolve for $\Delta t$ more, also evolving the existing reference state without any modification. Calculate their distance, re-normalize, and repeat until necessary. *Hint: be careful to evolve both initial conditions for exactly the same amount $\Delta t$!*

3.3 Compare your results from the previous two exercises with the result of `lyapunov` from DynamicalSystems.jl and confirm their correctness.

3.4 Implement the full algorithm for calculating the Lyapunov spectrum of a discrete dynamical system as discussed in detail in Sect. 3.2.2. Apply it to the standard map for $k$ ranging from 0.6 to 1.2, always starting with initial condition $(0.1, 0.11)$. Plot the resulting exponents. For a given $k$ what should be the sum of the two Lyapunov exponents? Use this answer to confirm the validity of your implementation.

3.5 Repeat the above exercise for a continuous dynamical system, specifically the Lorenz-63 system, (1.5), versus the parameter $\rho$ from 20 to 100, and plot the resulting three exponents. *Hint: since this is a continuous system you must make a "new" dynamic rule that contains (3.1) and plug that into your differential equations solver.*

3.6 Calculate analytically what the sum of the Lyapunov exponents of the Lorenz-63 should be for a given set of parameters, and use this fact to validate your implementation of the previous exercise.

3.7 Compare your Lyapunov spectrum results of the discrete and continuous systems of the previous exercises with the `lyapunovspectrum` function from DynamicalSystems.jl (or with Code 3.1, which is almost identical) and confirm again that they are correct.

3.8 Based on the discussed algorithm for calculating the Lyapunov spectrum, analytically show that for a 1D discrete dynamical system, its Lyapunov exponent is calculated by the expression

$$\lambda(x_0) = \lim_{n \to \infty} \frac{1}{n} \sum_{i=1}^{n} \ln \left| \frac{df}{dx}(x_i) \right|.$$

3.9 Calculate the Lyapunov spectrum of the towel map,

$$x_{n+1} = 3.8x_n(1 - x_n) - 0.05(y_n + 0.35)(1 - 2z_n)$$
$$y_{n+1} = 0.1((y_n + 0.35)(1 - 2z_n) - 1)(1 - 1.9x_n)$$
$$z_{n+1} = 3.78z_n(1 - z_n) + 0.2y_n$$

Based on the Lyapunov spectra of the towel map and the Lorenz-63 system, explain the movies in animations/3/volume_growth.

3.10 The maximum Lyapunov exponent is in principle approximated by the formula

$$\lambda_1 = \lim_{t \to \infty} \lim_{y_0 \to 0} \frac{1}{t} \ln \left( \frac{|\mathbf{y}(t)|}{|\mathbf{y_0}|} \right)$$

for a perturbation $\mathbf{y_0}$ of some initial condition $\mathbf{x_0}$. Consider a continuous dynamical system with a fixed point $\mathbf{x}^*$. If $\lambda_1$ is the largest Lyapunov exponent characterizing this "set" (a single point is also a set) and $\mu_i$ the eigenvalues of the Jacobian there, show that $\lambda_1 = \mathrm{Re}(\mu_1)$ with $\mu_1$ the eigenvalue with largest real part. *Hint: express the perturbation in the eigenvectors of the Jacobian as in Sect. 1.4 and then extract $\lambda_1$ via its definition.*

3.11 Repeat the above exercise for a discrete system to find that $\lambda_1 = \log(|\mu_1|)$ with $\mu_1$ the eigenvalue with largest absolute value.

3.12 Without using code, provide the signs of Lyapunov exponents (choosing between $+, 0, -$) the following kind of orbits should have, in the case of three dimensional discrete, or continuous, systems: stable fixed point, unstable fixed point, stable periodic orbit, stable quasiperiodic orbit, chaotic attractor. *Hint: for example, a stable fixed point of a discrete system is characterized by the triplet $(-, -, -)$.*

3.13 Modify the code you wrote in exercise 3.1/3.2 to provide a convergence time-series of $\lambda_1$ instead of the final value. For the Lorenz-63 system, report how the speed of convergence depends on the choice of $d_0, \Delta t$.

3.14 Modify the code you wrote in exercise 3.5 so that you can obtain timeseries of convergence for all Lyapunov exponents. How fast do the exponents converge for the Lorenz-63 system? How does convergence depend on the size of $\Delta t$?

3.15 Continuing from the above exercise, now apply the same process to the Hénon-Heiles system. Define a convergence measure and initialize several initial conditions all with the same energy. Produce a Poincaré surface of section where each point is color-coded by its convergence time (similarly with Fig. 3.7). What do you observe?

3.16 Consider the following 4D Hamiltonian system, defined by

$$H(\mathbf{q}, \mathbf{p}) = \frac{1}{2}\left(\mathbf{q}^2 + \mathbf{p}^2\right) + \frac{B}{\sqrt{2}}q_1\left(3q_2^2 - q_1^2\right) + \frac{1}{8}\mathbf{q}^4$$

with $\mathbf{q} = (q_1, q_2)$, $\mathbf{p} = (p_1, p_2)$ and $B = 0.5$. It is used in nuclear physics to study the quadrupole vibrations of the nuclear surface [54]. For it, create a figure similar to Fig. 3.7: generate Hamilton's equations of motion, decide on a reasonable energy value, generate initial conditions that all have the same energy, evolve them and obtain their PSOS, and color-code their PSOS plot according to their $\lambda_1$. *Hint: if you haven't solved exercise 3.2, you can use* lyapunov *from DynamicalSystems.jl.*

3.17 In Sect. 1.5 we discussed conservative and dissipative systems. Some systems are mixed in the sense of having sets in the state space that are either dissipative or conservative. A simple case of this is due to Sprott [55]

$$\dot{x} = y + 2xy + xz, \quad \dot{y} = 1 - 2x^2 + yz, \quad \dot{z} = x - x^2 - y^2. \tag{3.4}$$

For this system the initial condition $(1, 0, 0)$ leads to conservative motion on a 2-torus, while $(2, 0, 0)$ leads to dissipative motion on a chaotic attractor. Confirm that this is true by calculating how the volume of an infinitesimal sphere would change along the orbit in each case.

3.18  In Sect. 3.3 we mentioned that the "initial state $\mathbf{x}_0$ has to be located in the intersection $B_0 \cup \Phi^{-\Delta t}(B_1) \cup ... \cup \Phi^{-n\Delta t}(B_n)$ whose volume shrinks for increasing $n$". Why is it guaranteed that the volume of the intersection will shrink, and in fact will tend to 0 for $n \to \infty$, for chaotic dynamics?

3.19  Think of a dynamical system that converges to a periodic orbit or a fixed point. In such a case, why is it that we do not obtain information about the initial condition in the way it was discussed in Sect. 3.3?

# Chapter 4
# Bifurcations and Routes to Chaos

**Abstract** Dynamical systems typically depend on parameters. When these parameters change, the system may undergo qualitative changes in its behaviour. These changes are called *bifurcations*. Knowing about bifurcations means understanding how the system responds to parameter changes. In this chapter we review some standard bifurcation patterns, and discuss visualisation processes that allow one to detect (some) bifurcations. Identifying bifurcations numerically on a computer is not only possible, but also useful, so it is showcased here as well. We close the chapter with some of the so-called routes to chaos, characteristic ways one can transition from regular behaviour into chaotic behaviour by continuously changing a parameter of a system.

## 4.1 Bifurcations

When varying a parameter of a dynamical system, it is typically expected that the system's behavior will change smoothly (for example, the location of a fixed point changes smoothly with the parameter change). There are cases however, where a *smooth* change in one of the parameters of a dynamical system results in a significant, qualitative change in system behavior. For example, a previously stable fixed looses its stability or a new periodic orbit is "born". These scenarios of change are called *bifurcations*, and the parameter values at which bifurcations occur are called *critical* or *bifurcation points*.

Better understood via an example, we return to the simple 1D climate model we defined in (2.2). The visualizations there were done for constant parameter $\epsilon$. Let's now try to do the following: plot the fixed point(s) of the system while encoding their stability in the plot (i.e., the sign of $df/dx$) for various $\epsilon$ values (we have calculated the fixed points numerically here using a root finding algorithm, see Sect. 4.2.2). The result is shown in Fig. 4.1. See online animations/4/bifurcations_energybalance for an application that creates the figure interactively.

© The Author(s), under exclusive license to Springer Nature Switzerland AG 2022   53
G. Datseris and U. Parlitz, *Nonlinear Dynamics*, Undergraduate Lecture Notes in Physics,
https://doi.org/10.1007/978-3-030-91032-7_4

**Fig. 4.1** Fixed points $T^*$ of
(2.2) versus $\epsilon$. Two
half-filled points indicate the
two saddle-node bifurcations
that occur. The purple and
cyan arrows highlight the
hysteresis that arises in the
system

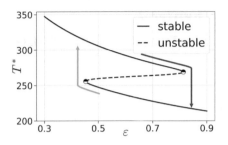

Figure 4.1 is called a *bifurcation diagram* because it displays how the locations
of the fixed points, and their stability, change with a change in a parameter. Starting
from small $\epsilon = 0.3$, only one fixed point exists, and its value is changed smoothly
as $\epsilon$ changes. At some critical parameter value $\epsilon_1 \approx 0.45$ two new fixed points are
created, as a pair of stable-unstable points. The same process happens again at around
$\epsilon_2 \approx 0.8$ but now in reverse: a stable and an unstable point "annihilate" each other.

### 4.1.1  Hysteresis

Let's imagine for a moment that we set $\epsilon = 0.5$ and the system is at the fixed point
of the "bottom branch" at $T^* \approx 236$, see Fig. 4.1. We then continuously and slowly
decrease $\epsilon$. The system always has time to equilibrate at the new nearby fixed point,
until we reach the bifurcation point at around $\epsilon \approx 0.45$ (follow the cyan arrow in
Fig. 4.1). What will happen once we cross the bifurcation threshold, $\epsilon < 0.45$? Well,
the system will go to the only available fixed point at the "top branch", at $T^* \approx 315$.
This fixed point now became *globally* attractive and is "far away" from the fixed
point where the system was just before. Although this looks like a "jump" in the
diagram, the transition takes some finite transient time. Sometimes such events are
also referred to as catastrophes in the literature.

The interesting part happens once we try to reverse the change of $\epsilon$, i.e., increase
it. Even though the previously existing stable fixed point of the lower branch comes
into existence again, the system does not go there, as there is already a stable fixed
point nearby its current state. Thus, the system will now follow the purple arrow,
until we cross yet another bifurcation at $\epsilon \approx 0.8$, where the system will once again
abruptly switch "branches". This phenomenon is called *hysteresis* and may have
important practical implications.

In the context of the simple model for the temperature on Earth it means that once
a transition to the upper temperature branch has taken place due to a decrease of
the effective emissivity $\varepsilon$ (greenhouse effect), you cannot undo this by increasing $\varepsilon$
again a bit. Instead, you would have to increase the emissivity *significantly*, such that
the bifurcation and transition at the larger value of $\varepsilon$ occurs and the Earth would go
back to lower temperatures. These scenarios are also called *tipping*, and bifurcations

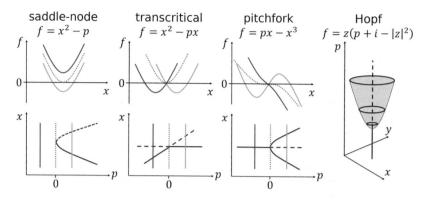

**Fig. 4.2** Typical named bifurcations, along with functional forms $f$ that result in them for one dimensional continuous dynamics $\dot{x} = f(x)$ with parameter $p$. In the functional forms, dotted orange curves correspond to the parameter value at which the bifurcation takes place. In the bifurcation plots (bottom row), stable components are solid lines, unstable are dashed lines. For the Hopf bifurcation, $z \in \mathbb{C}$ and $i$ is the imaginary unit

are one way that leads to tipping. In Chap. 12 we will discuss different tipping mechanisms that can occur for many ecological, economical, or other systems.

## 4.1.2  Local Bifurcations in Continuous Dynamics

In the previous section we presented two variants of the *saddle-node* bifurcation. The name means that out of nothingness a node and a saddle point (also called hyperbolic fixed point) are created. Some bifurcations are "typical", in the sense that they occur frequently in different dynamical systems. Many bifurcations have in common that they take place at some (fixed) point in state space and they are therefore called *local bifurcations*. Figure 4.2 outlines these bifurcations for continuous systems. Information about the flow around the bifurcation point is sufficient to completely characterize local bifurcations.

With the *transcritical bifurcation*, two existing fixed points exchange their stability at the bifurcation point. In a *pitchfork bifurcation*, an existing fixed point gives rise to two new fixed points with the same stability property, while the original one continues to exist but with changed stability (i.e., if it was stable it becomes unstable). In case of the *Hopf bifurcation* shown in Fig. 4.2, a stable fixed point becomes unstable, and a limit cycle comes into existence. This scenario requires a at least two-dimensional dynamics.

In Fig. 4.2 we show bifurcations in one or two-dimensional continuous systems and we also display the corresponding *normal forms*, simple dynamical systems $\dot{x} = f(x)$ that result in the respective bifurcations. Do not be misled by the simplicity of the normal forms however, because these typical bifurcations can occur in higher dimensional systems just as well. Local bifurcations can be "detected" by looking at the Jacobian eigenvalues at the fixed point. At a bifurcation, at least one of the eigenvalues obtains a vanishing real part. For example, in the first three bifurcations presented the Jacobian eigenvalue (i.e., the slope of $f$) becomes zero at the bifurcation point $p = 0$. In the case of the Hopf bifurcation, both Jacobian eigenvalues of the 2D system have vanishing real parts at the bifurcation point. Exactly after the bifurcation point they obtain a positive real part (unstable) but also a non-zero imaginary part, which gives rise to the stable limit cycle.

Another important point is that for all the bifurcations shown, one can create an alternative bifurcation "version" where the stability and instability of the sets is exchanged. The "direction" of the bifurcation with respect to the $p$-axis can also be reversed, depending on the specific form of $f$. Thus, another possibility for, e.g., the Hopf bifurcation is that an unstable fixed point becomes stable and an unstable limit cycle is created. In the literature you will find the terms *subcritical* and *supercritical* to refer to the two possible versions of each bifurcation.

### 4.1.3   Local Bifurcations in Discrete Dynamics

The bifurcations illustrated in Fig. 4.2 can also occur in discrete systems. The saddle-node bifurcation is sometimes called a *tangent* bifurcation for discrete systems, and the Hopf bifurcation is slightly different, as instead of a limit cycle a periodic or quasiperiodic orbit emerges from the fixed point. In addition there is a bifurcation in 1D discrete systems that has no counterpart in 1D continuous systems: the period doubling bifurcation, where a period-m orbit becomes unstable and gives rise to a stable period-2m orbit. To illustrate this type of bifurcation we use the logistic map $x_{n+1} = f(x_n) = rx_n(1 - x_n)$. For $1 \leq r < 3$ the map has a stable fixed point $x^* = (r - 1)/r$. At $r = 3$ the local slope[1] at the fixed point is $-1$ and a period-doubling bifurcation takes place: the fixed point $x^*$ becomes unstable and a stable period-2 orbit emerges as illustrated in Fig. 4.3. As you will prove in the exercises, a stable period-2 orbit of a map $x_{n+1} = f(x_n)$ is in fact a stable fixed point of the map $x_{n+1} = f^{(2)}(x_n) = f(f(x_n))$, called the *second iterate* of $f$.

In Fig. 4.3 we illustrate the period-doubling bifurcation using a *cobweb diagram*. Cobweb diagrams are a simple way to visualize 1D discrete system dynamics. In Fig. 4.3a the purple curve is $f$, while the black lines follow the evolution of $x_n$. Starting from an $x_n$ anywhere on the $x$ axis, we go up (or down as necessary) until we reach the $f$ curve. This provides $x_{n+1}$. We then go right (or left) until we cross the diagonal. This maps $x_{n+1}$ to the x-axis, as we need it as the new seed to generate the next step. After reaching the diagonal we again simply go up (or down) until reaching $f$, which provides $x_{n+2}$, and repeat the process for how many steps we need. Notice that intersections of $f$ with the diagonal are fixed points of the map,

---

[1] Recall from Sect. 1.4 that for discrete systems we check stability by comparing absolute value of eigenvalues with 1: $|\mu| < 1$.

**Fig. 4.3** Cobweb diagrams
for the logistic map for
$r = 2.8 < 3$ (fixed point)
and $r = 3.3 > 3$ (period-2)

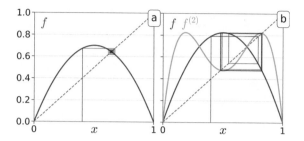

since by definition they satisfy $f(x) = x$. In Fig. 4.3b the trajectory converges to a
square on the cobweb diagram, whose corners are where the curve $f^{(2)}$ intersects the
diagonal, i.e., the fixed points of $f^{(2)}$. See also animations/4/logistic_cobweb online.

The local bifurcations discussed here occur when varying a single parameter of
the system. They are therefore called *co-dimension 1* bifurcations and constitute the
class of simplest bifurcations. Bifurcations are said to have *co-dimension k* if it is
necessary to (simultaneously) vary $k$ parameters for the bifurcation to occur and to
describe it. A co-dimension 2 bifurcation, for example, occurs at an isolated point in
a two-dimensional parameter plane.

### 4.1.4 Global Bifurcations

Bifurcations are not only about just altering stability or existence of fixed points, but
also cover other scenarios where extended invariant sets (e.g., limit cycles) collide
with other sets and the resulting changes in the topology of the trajectories are not
restricted to a small neighbourhood, but have a global impact. Hence the name *global
bifurcation*. Examples are the *homoclinic bifurcation* in which a limit cycle collides
with a saddle point and then vanishes.

If a chaotic attractor is involved, global bifurcations are often called *crisis*. A
typical example is a chaotic attractor which collides with its own basin boundary
upon parameter variation and "runs out", i.e., the trajectory goes to another attractor
in state space. This *boundary* or *exterior crisis* is illustrated in Fig. 4.4 for the Hénon
map. Figure 4.4a shows the Hénon attractor (black) and its basin of attraction (light
yellow) for $a = 1.4$. The boundary is defined by the stable manifold of the saddle
point of the Hénon map. Parts of the attractor are very close to the basin boundary.
If the parameter $a$ is slightly increased the attractor collides with the boundary of its
basin and the chaotic attractor turns into a chaotic saddle. The difference between a
chaotic saddle and a chaotic attractor is similar to the difference between a saddle
point and an attractive fixed point (see Fig. 2.2). A chaotic saddle is a complicated
(fractal) invariant set, like a chaotic attractor, however in the former all orbits that
are nearby (but not directly on) the chaotic saddle will eventually *diverge* from it.

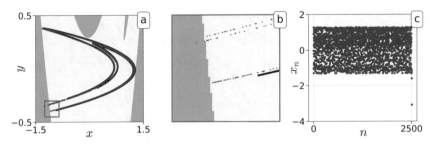

**Fig. 4.4** Boundary crisis of the Hénon attractor. **a** Chaotic attractor (black) and its basin of attraction (light yellow) for $a = 1.4$ and $b = 0.3$. Initial conditions from the region plotted in cyan diverge to $-\infty$. **b** Zoom-in of (**a**) for the indicated square. **c** Timeseries of $x_n$ for $a = 1.427$ and $b = 0.3$ and initial condition $(0.1, 0.1)$. After a chaotic transient the iterations quickly diverge to $-\infty$. The trajectory is also plotted in (**a**, **b**) in purple

While near the chaotic saddle, orbits sooner or later will fall into a region "beyond the former attractor basin" and then start to diverge (as shown in Fig. 4.4b, c). Such a *chaotic transient* can take very long and therefore, with finite time simulations or measurements, it can be challenging to find out whether a systems really possesses a chaotic attractor (with *persistent chaos*) or whether we observe "just" *transient chaos*. Another type of global bifurcation is the *interior crisis*, where the size of the chaotic attractor suddenly increases due to a collision with an unstable fixed point or an unstable periodic orbit inside the basin of attraction. And if two chaotic attractors merge to a single attractor at some critical parameter value, the corresponding global bifurcation is called an *attractor merging crisis*.

## 4.2   Numerically Identifying Bifurcations

### 4.2.1   Orbit Diagrams

The first way to numerically identify a bifurcation is to identify qualitative changes in the asymptotic dynamics of the system, versus a change in a parameter $p$. Orbit diagrams are a simple way to visualize this change for varying $p$ for both discrete and continuous systems. For discrete systems, you choose an initial condition $\mathbf{x}_0$ and a (start) value of the parameter $p$, evolve the system for a transient amount of steps $n_0$, and then plot one of the system variables $x_j$ for $n$ steps. Then you increment or decrement $p$ by a small amount $\delta p$ and repeat the iteration, either using the same initial value $\mathbf{x}_0$ or using the last result from the previous computation with $p$.[2] Repeat

---

[2] Often the last state of a previous parameter value is used and the parameter value is first increased in small steps and then decreased. In this way hysteresis scenarios can be detected (like in Fig. 4.1). Another option is to search for coexisting attractors using many different initial conditions for each parameter value.

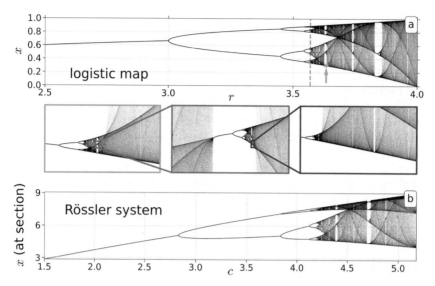

**Fig. 4.5** Orbit diagrams for **a** the logistic map ($x_{n+1} = rx_n(1 - x_n)$) and **b** the Rössler system ((4.1), $a = 0.2, b = 0.2$). All points are plotted with some transparency and thus darker color means higher density of points. Middle row are successive zoom-ins of panel a, starting with the small cyan rectangle (pointed by the arrow). The value of $r_\infty$ is denoted with a vertical dashed gray line. The function `interactive_orbitdiagram` from InteractiveDynamics.jl allows one to interactively explore and zoom into orbit diagrams, see also animations/4/orbitdiagram online for a recorded video

this for many values of $p$, and plot the resulting values of the recorded state variable $x_j$ versus $p$. An example is shown in Fig. 4.5a. Orbit diagrams can be measured and plotted also for real world (experimental) systems. Since in experiments initial values can rarely be manipulated the last state before changing the parameter is in most cases the initial value for the next step when slowly varying a control parameter.

Here is how to interpret Fig. 4.5a. We see that for $r$ from 2.5 to 3.0 only a single point is visible. Given that the system is discrete, this means that the system resides at a stable fixed point and all plotted values of the system variable are the same. At $r = 3.0$ we see a "split" and there are two branches. This means that a period-doubling bifurcation has taken place and the system now has a period of length 2, instead of a fixed point, and the plotted variable can take two values, while the previous fixed point (for $r < 3$) became unstable. At around $r \approx 3.449$ four branches emerge, which means that the system has a (stable) period-4 orbit. This period-4 attractor is the result of a period-doubling bifurcation of the stable period-2 orbit which looses its stability at $r \approx 3.449$.

At other points of the orbit diagram (e.g., look at $r = 4$) an entire range of the real numbers is covered by the recorded variable. This means that the long-term dynamics of the system does not settle in a stable periodic orbit with finite number of points, indicating either chaotic or quasiperiodic motion (as both continuously fill spans of the real line for the recorded variable). Calculating, e.g., Lyapunov exponents (see

Chap. 3) can help to distinguish between the two cases. In the case of the logistic map shown here, quasiperiodic dynamics does not occur and the dynamics for $r = 4$ is chaotic.

For continuous systems the process is slightly different. That is because it doesn't make much sense to record the state of a continuous system since, e.g., a simple periodic orbit covers a full range of real numbers (while for discrete systems a periodic orbit is by definition a finite set of numbers). This is an excellent opportunity to use the Poincaré surface of section (Sect. 1.6) which will turn the continuous system into a discrete one. We therefore record some variable in the Poincaré section of the system at a predefined hyper-plane defined by $y = 0$ and $\dot{y} < 0$ (see Fig. 4.7a). As before, we first evolve the system transiently and then start plotting these states versus $p$. This is how Fig. 4.5b was produced, using the Rössler system

$$\dot{x} = -y - z, \quad \dot{y} = x + ay, \quad \dot{z} = b + z(x - c) \qquad (4.1)$$

as an example, which has a chaotic attractor for $a = 0.2, b = 0.2, c = 5.7$, shown in Fig. 4.7a. See Code. 4.1 as well.

For parameter $c = 2$ we have a single point in the orbit diagram, which means a single point in the Poincaré section. This in turn means a stable periodic orbit in the full state space (see Sect. 1.6). When the branch splits into two, e.g., at $c = 3$ we have a periodic orbit that crosses the Poincaré section at two distinct points. The orbit in the state space of the continuous system is still closed, but it must "loop around" two times. If the orbit intersects the Poincaré section (in a given direction) at $m$ points, it is called a period-$m$ orbit, borrowing the terminology of the discrete systems. Just like in the discrete case a continuously filled span of plotted values indicates either chaotic or quasiperiodic motion.

**Code 4.1** Computing the orbit diagram of the Rössler system, Fig. 4.5b

```
using DynamicalSystems, PyPlot
ro = Systems.roessler(); u0 = ro.u0
cs = 1.5:0.01:5.2
plane = (2, 0.0) # hyperplane for the Poincare section
for c ∈ cs
    set_parameter!(ro, 3, c)
    psos = poincaresos(ro, plane, 10000; Ttr = 2000, u0)
    plot(fill(c, length(psos)), psos[:, 1]; lw = 0,
    marker = "o", ms = 0.2, alpha = 0.1)
    u0 = psos[end]
end
```

### 4.2.2 Bifurcation Diagrams

Orbit diagrams (Fig. 4.5) do not show repelling sets (e.g., unstable fixed points), because the long term dynamics of the system will never stay there. Bifurcation diagrams (Fig. 4.2) are curves that provide the fixed points and their stability for a given parameter value. In contrast to orbit diagrams they show both stable and unstable fixed points (and in some cases also periodic orbits), but cannot show chaos or quasiperiodicity. Therefore we distinguish between the two types of diagrams, although in the literature orbit diagrams are often also called bifurcation diagrams.

While sometimes it is possible to compute bifurcation diagrams analytically, because one can extract the fixed points of the system and their stability analytically, this becomes harder or even impossible in higher dimensional systems. Therefore, in general you need a numerical method that can detect and track not only stable but also unstable fixed points. For locating fixed points of discrete systems you can consider the fixed point problem $f(\mathbf{x}) = \mathbf{x}$ as a root finding problem for the function $g(\mathbf{x}) = f(\mathbf{x}) - \mathbf{x}$ and use some numerical algorithm to locate the root(s) $\mathbf{x}^*$ with $g(\mathbf{x}^*) = 0$. Similarly, for continuous systems you may choose $g = f$, because $g(\mathbf{x}^*) = f(\mathbf{x}^*) = \dot{\mathbf{x}}^* = 0$.

In both cases a solution can be computed iteratively using *Newton's method*

$$\mathbf{x}_{j+1} = \mathbf{x}_j - \delta_j J_g^{-1}(\mathbf{x}_j) g(\mathbf{x}_j) \tag{4.2}$$

where $j$ counts the iterations, $\mathbf{x}_0$ is the initial value, $J_g^{-1}(\mathbf{x}_j)$ the inverse of the Jacobian matrix of $g$ and $\delta_j$ a damping factor that can be chosen $\delta_j = 1$ if $\mathbf{x}_j$ is already close to the root $\mathbf{x}^*$ of $g$. To make sure that the iterations converge to $\mathbf{x}^*$ if you start further away from it you could check whether the current iteration step $j$ brings you closer to the zero, i.e., whether $\|g(\mathbf{x}_{j+1})\| < \|g(\mathbf{x}_j)\|$. If this is not the case you have to go back to $\mathbf{x}_j$ and try again with a smaller step size by reducing the damping factor (for example, replacing $\delta_j$ by $\delta_j/2$). Such a *damped Newton's method* does not care about the stability of the fixed point of $f$, it converges for both stable and unstable fixed points. One then can, in principle, densely populate the mixed space $(\mathbf{x}, p)$ with several initial conditions $(\mathbf{x}_0, p_0)$ and use Newton's method to each one, until all have converged to some stable/unstable fixed point, thus creating all bifurcation curves as in Fig. 4.1 (typical dynamical systems have more than one bifurcation curve).

### 4.2.3 Continuation of Bifurcation Curves

For high dimensional systems the aforementioned approach is impractical, because it would take too many initial conditions to fill the state space and thus too much time to do the computation. What we want to do instead is somehow track how fixed points change once changing $p$. This is called a *continuation*, and it relies on a two-step

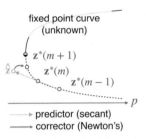

**Fig. 4.6** Illustration of bifurcation curve continuation via a predictor-corrector process. $\mathbf{z} = (\mathbf{x}, p)$ is the mixed state-parameter vector. A new vector $\hat{\mathbf{z}}$ is predicted via the secant method, which is then corrected into the fixed point vector $\mathbf{z}^*$ by the corrector (Newton's method). A saddle node bifurcation is also indicated in the figure as a half-filled circle (dashed line corresponds to unstable fixed point)

process: a *predictor* that provides a new initial guess, and a *corrector* that corrects this guess to the (nearest) fixed point, as illustrated in Fig. 4.6. Here we will present the simplest possible set of a predictor-corrector (that works well nevertheless), although much more advanced algorithms can be found in the literature.

For clarity of notation, let $\mathbf{z} = (\mathbf{x}, p)$ be the $(D + 1)$-dimensional vector in the *mixed state-parameter space*, and $\mathbf{z}^*(m) = (\mathbf{x}^*(p_m), p_m)$ representing the $m$-th fixed point we have found during the process (throughout this section $m$ enumerates the found points along the bifurcation curve). The simplest predictor, called the *secant predictor*, yields a new estimate $\hat{\mathbf{z}}$ as a linear extrapolation of the two[3] previously found points, i.e., $\hat{\mathbf{z}} = \mathbf{z}^*(m) + \Delta\mathbf{z}^*(m)$, $\Delta\mathbf{z}^*(m) = \mathbf{z}^*(m) - \mathbf{z}^*(m - 1)$. Then $\hat{\mathbf{z}}$ can be used as an initial condition for Newton's method which is iterated until it converges to $\mathbf{z}^*(m + 1)$.

But there is a problem. As we have seen before in Figs. 4.1 and 4.2 there may be, for example, saddle node bifurcations where the curve does not change in the direction of the $p$-axis but has a tangent which is perpendicular to it. These cases are called *turning points*, and an example is illustrated in Fig. 4.6. If our prediction for $p$ already brings us beyond the turning point, then Newton's method, as defined in (4.2), will fail, as there is no fixed point to converge to for this $p$ value. That is why we need a better *corrector*, which has the possibility of not only changing $\mathbf{x}$, but also $p$. Instead of $g(\mathbf{x})$, we now consider $h(\mathbf{z}) = g(\mathbf{x}, p)$ as a function of $D + 1$ variables. Remember, the output of $g$ is a vector of length $D$, as $g$ is either $f$ for continuous systems or $g(\mathbf{x}, p) = f(\mathbf{x}, p) - \mathbf{x}$ for discrete. To keep the notation clear, we define a new function $h$, which is currently identical to $g$, but will change in a moment (see below). In the mixed state-parameter space, the Jacobian $J_h$ of this $h$ has $D + 1$ columns, as the last extra column is the derivative of $g$ with respect to $p$.

The difficulty now is that we have a set of $D$ equations (the elements of $h(\mathbf{z})$) for $D + 1$ unknowns (the elements of $\mathbf{x}$ and the extra $p$). The simplest way to resolve

---

[3] To start the algorithm you have to provide a first initial guess $(\mathbf{x}_0, p_0)$. Once this guess converges to $(\mathbf{x}_0^*(p_0), p_0)$, you also provide a first $\Delta\mathbf{z}_0$ to kick-start the secant method.

this is to expand $h$, so that it has a $(D + 1)$-th element. To do this we select one of the elements of $\mathbf{z}$, say the $k$-th, and demand that it takes some specified value $\tilde{z}_k$ in the next continuation step $m + 1$, i.e., $\mathbf{z}_k(m + 1)$ should equal $\tilde{z}_k$. To incorporate this demand into Newton's method, we only have to define the $(D + 1)$-th element of $h$ as $h_{D+1}(\mathbf{z}) = z_k - \tilde{z}_k$, as the "zero" of this entry (which is what Newton's method finds) satisfies by definition our demand. At each continuation step $m$ it is recommended to choose as $k$ the index of $\Delta\mathbf{z}^*(m)$ that has the largest magnitude, which is the index of the variable that has changed the most in the previous continuation step. Now, to recap, at each continuation step $m$ we have defined $h$ and its Jacobian as

$$h(\mathbf{z}) = (g_1(\mathbf{z}), \ldots, g_D(\mathbf{z}), z_k - \tilde{z}_k), \quad J_h(\mathbf{z}) = \begin{bmatrix} \dfrac{\partial g_1}{\partial x_1} & \cdots & \dfrac{\partial g_1}{\partial x_D} & \dfrac{\partial g_1}{\partial p} \\ \vdots & \ddots & \vdots & \vdots \\ \dfrac{\partial g_D}{\partial x_1} & \cdots & \dfrac{\partial g_D}{\partial x_D} & \dfrac{\partial g_D}{\partial p} \\ \delta_{1,k} & \cdots & \cdots & \delta_{D+1,k} \end{bmatrix} \tag{4.3}$$

where in the last row of $J_h$ the $k$-th element is 1 and all other elements are 0 ($\delta_{i,j}$ is the Kronecker $\delta$). Plugging this $h$ and $J_h$ into the Newton's method of (4.2) our initial guess will correctly converge to the nearest fixed point, even in the case that we have over-shooted a turning point. Notice that even with this approach one should start the whole continuation process with several initial conditions, so that all bifurcation curves are found, as typical systems have more than one such curve.

For each bifurcation curve identifying the actual bifurcation (i.e., when the stability of a fixed point changes) is easy. Based on the eigenvalues of the Jacobian $J_f$ at each point of the curve, we can tell whether a fixed point is stable or unstable. If the stability of a fixed point changes as we stroll through the curve, then we have identified a (local) bifurcation point. If we also find that once the fixed point becomes unstable its eigenvalues obtain non-zero imaginary part, then we have detected a Hopf bifurcation.

## 4.3 Some Universal Routes to Chaos

### 4.3.1 Period Doubling

Alright, let's talk about the elephant in the room... Why is Fig. 4.5 so ridiculously complicated, and how can it be that the bottom and top panels look so similar, even though one is a 1-dimensional discrete system and the other a 3-dimensional continuous system?! Well, Fig. 4.5 showcases the period doubling route to chaos, where a (discrete) system starts with a fixed point, then successively splits into period 2, then 4, then 8, and the periodicity continues to double, until we reach period $\infty$. And then something even crazier happens, but we'll get to that later.

In Sect. 4.1.2 we have already discussed the period doubling bifurcation, using the logistic map as an example. It turns out that for the logistic map such period doubling bifurcations continue to occur at higher order iterates of the map as we increase the parameter $r$ further: at a parameter $r_k$ a stable $2^{k-1}$-period orbit becomes unstable, and a new stable $2^k$-period orbit is born. Equivalently, we can say that a fixed point of the $f^{(2^{k-1})}$ iterate of $f$ undergoes a period doubling bifurcation.[4] This first happens at $r_1 = 3$, where initially we had period $2^0$ (fixed point), that breaks into a period $2^1$. The same story repeats ad infinitum for all $k$ at parameter $r_k > r_{k-1}$.

Importantly, each successive period doubling happens geometrically faster with respect to increasing $r$, meaning that

$$\frac{|r_{k+1} - r_k|}{|r_{k+2} - r_{k+1}|} \xrightarrow{k \to \infty} \delta \approx 4.669201609... \quad \Rightarrow \quad |r_\infty - r_k| \approx c \cdot \delta^{-k} \qquad (4.4)$$

with $\delta$ the so-called *Feigenbaum constant* and $c$ some constant. As a result $r_\infty$ is finite, i.e., infinite period doublings happen at finite parameter increase, and we thus reach a state of infinite periodicity at $r_\infty \approx 3.56995...$ (at $r_\infty$ the Lyapunov exponent $\lambda$ is zero and becomes positive for $r > r_\infty$).

And what is going on after $r_\infty$? The successive zoom-ins of Fig. 4.5 reveal that this process of period doubling repeats itself at arbitrary small scales, starting with a period of arbitrary length. In fact, this orbit diagram is a fractal (see Chap. 5), because it is self-similar: it repeats itself at arbitrarily smaller scales, *forever*. This means that for $r > r_\infty$ there are *infinite* periodic "windows" with period doubling sequences each followed by its own chaotic dynamics.

Astonishingly, this period-doubling orbit diagram is universal. It turns out that the same orbit diagram exists for all 1D maps where $f$ is a uni-modal function.[5] When we say same, we don't mean "similar". Quite literally there is quantitative universality. For example, all period doublings are governed by the same constant $\delta$ mentioned above. Thus, the precise form of $f$ is more or less inconsequential, provided it satisfied the aforementioned properties.

This universality doesn't say much on why the orbit diagram of the Rössler system behaves the same way. In fact, many continuous systems (e.g., also the Lorenz-63) display the same period doubling. We can see how by utilizing the same Poincaré section that we used to produce the orbit diagram. In Fig. 4.7 we illustrate firstly how the Poincaré section was obtained. Remember, the Poincaré section, which in essence results in a discrete dynamical system, contains all information of the

---

[4]  Notice that a period-$j$ orbit is in fact a fixed point of the dynamical system $x_{n+1} = f^{(j)}(x_n) = f(f(f \ldots (x_n) \ldots))$ (also called the $j$-th iterate of $f$). A stable/unstable fixed point of $f^{(j)}$ is also a stable/unstable fixed point of $f^{(j \cdot \ell)}$ for any $j, \ell \in \mathbb{N}$.

[5] The Feigenbaum constant is universal for 1D maps with a negative Schwarzian derivative $\frac{f'''(x)}{f'(x)} - \frac{3}{2}\left(\frac{f''(x)}{f'(x)}\right) < 0$ in a bounded interval. The value $\delta \approx 4.669201609...$ holds for all maps with a quadratic maximum. In general, the order of the leading terms of the Taylor expansion at the maximum determines the universality class the map belongs to and the value of $\delta$.

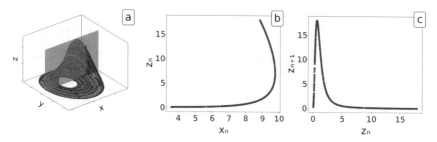

**Fig. 4.7  a** A trajectory of the Rössler system with the plane used for the Poincaré section (actual section in purple). **b** The Poincaré section. **c** Its Lorenz map

continuous case as well because of the uniqueness of solutions of the differential equations (Sect. 1.1.2).

Now we produce what is called the *Lorenz map*: we simply plot the successive entries of $z_n$ versus the previous ones $z_{n-1}$. In general it is not possible to express $z_{n+1} = f(z_n)$, because both $x_n, z_n$ are needed to produce $z_{n+1}$ (the Poincaré map is two-dimensional). Here, however, we see something special: The set of points in Fig. 4.7c resembles a curve, meaning that it is indeed possible to express $z_n \to z_{n+1}$ approximately as a 1D discrete dynamical system, $z_{n+1} \approx f(z_n)$. This is possible for the Rössler system (and other low-dimensional continuous systems) because its chaotic attractor is locally very flat and approximately two-dimensional (we will discuss this more in Sect. 5.3.2 when talking about the fractal dimension). Regardless, we can now understand why the Rössler system displays the same period doubling route to chaos as the logistic map: the function $f$ approximately describing $z_n \to z_{n+1}$ is unimodal! So the same type of dynamics and underlying universality of unimodal maps applies to the sequences of $z_n$. This in turn propagates to the Poincaré map $(x_n, z_n)$ since $z_n$ and $x_n$ are approximately connected by a one dimensional curve, and thus one can express approximately $x_n$ as a function of $z_n$. Because of uniqueness, this then finally also propagates to the 3D continuous dynamics!

### 4.3.2  Intermittency

Intermittency[6] is a characteristic type of chaotic behavior which, in some sense, can also be considered as a route to chaos. The time evolution of a dynamical system is called intermittent when for long periods of time it follows approximately regular dynamics, called *laminar phases*, which are disrupted by shorter irregular phases, called *chaotic bursts*. An example of intermittent timeseries is shown in Fig. 4.8b, c, where a chaotic burst with length $c$ and a laminar phase with length $\ell$ are annotated.

---

[6] You can observe intermittency by setting $r$ slightly below $1 + \sqrt{8}$ in the interactive cobweb application for the logistic map. The laminar phases will be period-3 segments. See animations/4/logistic_intermittency online.

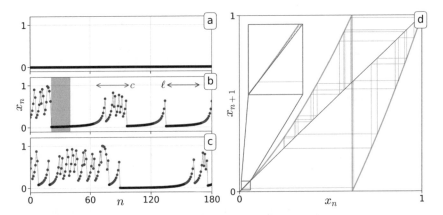

**Fig. 4.8  a, b, c** Timeseries of the map of (4.5) for $\varepsilon = -0.1, 0.0001, 0.05$. Chaotic bursts are indicated with purple color. **d** Cobweb diagram for the trajectory of (**b**) with function $f$ plotted in cyan. The inset is the part of the trajectory in the orange rectangle, and highlights how intermittency arises

Besides the fact that intermittency by itself is chaotic motion, the sequence of lengths of laminar phases and chaotic bursts is also chaotic.

The core of why intermittency happens can be understood without going into mathematical details. The key ingredient of the story is *trapping*. In the state space of a dynamical system that displays intermittency, there exists some structure around which the dynamics displays approximately regular dynamics (recall that regular dynamics are fixed points, periodic, or quasiperiodic motion). Importantly, the trajectory gets *trapped*, or *sticks*, in this structure for long times, until it escapes again. The long trapped phases are the laminar phases, where the trajectory follows approximately the regular dynamics of the structure it has been trapped in. The escape results in a chaotic burst.

What this trapping structure can be, and how the trapping mechanism works, depends a lot on the system and the scenario of intermittency being displayed, since different mechanisms can lead to intermittency. The simplest possible scenario to illustrate intermittency is from a 1D discrete map

$$x_{n+1} = f(x_n) = [(1 + \varepsilon)x_n + (1 - \varepsilon)x_n^2] \mod 1 \qquad (4.5)$$

which for $\varepsilon < 0$ has an attracting fixed point $x^* = 0$ (Fig. 4.8a) while for $\varepsilon > 0$ displays intermittency (Fig. 4.8b, c). Figure 4.8d presents a cobweb diagram of this map which elucidates what is going on. For $\varepsilon > 0$ and near $x = 0$ the function $f$ is really close to the diagonal line $f(x) = x$ without crossing it (see inset). Therefore the trajectory gets "stuck" there, making the dynamics look like that of a stable fixed point for long times.

The motivation for calling this behavior a route to chaos comes from the fact that for $\varepsilon < 0$ we have a fixed point, while for $\varepsilon > 0$ the trajectory gets stuck near the remnant of the fixed point. Increasing $\varepsilon$ makes the trajectory more and more chaotic (the maximum Lyapunov exponent increases with $\varepsilon$), while the laminar phases become shorter and less frequent. However, it is important to understand that *any* amount of intermittency is deterministic chaos: the transition to chaos is sharp, not gradual.

Some characteristic properties accompany intermittent timeseries in most typical cases. First, the power spectra display the so-called "$1/f$-noise" (with $f$ standing for frequency). This simply means that the power spectral density $S$ decays with frequency as a negative power law, $S \sim f^{-\alpha}$, $\alpha \in (0.5, 1.5)$. As a result, the auto-correlation function of the timeseries becomes a power law as well. Second, the probability distribution of the length of the laminar phases (called laminar lengths) is also typically a power law, $P(\ell) \sim \ell^{-\beta}$.

# Further Reading

An in-depth treatment on bifurcations in general can be found in *An Exploration of Dynamical Systems and Chaos* by Argyris et al. [5], or in *Elements of Applied Bifurcation Theory* by Kuznetsov [56]. See also [57] by Haragus and Loss for a discussion of local bifurcations in high dimensional dynamics. Software that can apply the analysis we described in Sect. 4.2.2 are, e.g., BifurcationKit.jl, MatCont, AUTO or COCO among others, and they typically implement much more elaborate algorithms than what described here. Good textbooks on numerically computing bifurcations and the continuation process of Sect. 4.2.2 are *Numerical Continuation Methods* by Allgower and Georg [58] and *Recipes for Continuation* by Dankowicz and Schilder [59].

Hysteresis is Greek for "lagging behind". It was first discussed in the context of magnetism. Notice that there the quantity that displays hysteresis is not the state of the system but an aggregated property (average alignment of magnetic moments) of the components of the system (since magnetism is better understood with the theory of statistical physics and not of dynamical systems).

The period doubling route to chaos and its universality was rigorously presented in the work of Feigenbaum [60] and also Coullet and Tresser [61] and Grossmann and Thomae [62]. A collection of early references, and most related theorems, is provided in the Scholarpedia entry [63]. Arguably the most influential usage of the logistic map and the period doubling route to chaos is [6] by May. May's goal was to demonstrate that incredibly complex behavior can arise from incredibly simple rules, provided they are nonlinear. Period doubling has been measured many times in real world systems [64, 65].

Pomeau and Manneville [66, 67] were the first to discuss intermittency in dynamical systems and in appropriate mathematical depth, and generated a map (discrete system) that can be used to model and understand intermittency. The map of (4.5) is in fact a modification of the Pomeau-Manneville map [66]. Intermittency is a common phenomenon in nonlinear dynamics. In systems that display period doubling universality in their orbit diagram, intermittency almost always appears at the end of chaotic windows, just before a period window opens up. But intermittency also appears in several other classes of systems, and even billiards (Chap. 8, [68]). Many real world systems display intermittency, like electronic circuits, fluid flows and convection, among others [69].

The routes to chaos that we discussed here are mathematically rigorous. *Deterministic Chaos* by Schuster and Just [69] provides this rigor and explains in detail how universality emerges using renormalization theory (and also gives further historical references).

## Selected Exercises

4.1 For the 1D continuous system $\dot{x} = \mu + 2x - x^2$, find the critical value of the parameter $\mu$ at that a bifurcation occurs and describe the bifurcation.

4.2 Use Newton's method and the continuation described in Sect. 4.2.2 to reproduce the bifurcation diagram of the 1D climate model, (2.2) and Fig. 4.1.

4.3 Consider the system $\dot{x} = f(x; p, \epsilon) = \epsilon + px - x^3$. For $\epsilon = 0$ it is the normal form of the pitchfork bifurcation shown in Fig. 4.2. Produce bifurcation diagrams of the system versus $p$ for various $\epsilon$ values. Animate the "evolution" of the bifurcation diagram when increasing $\epsilon$ from $\epsilon < 0$ to $\epsilon > 0$. What do you observe?

4.4 The van der Pol oscillator  is defined by

$$\frac{d^2x}{dt^2} + \mu(1 - x^2)\frac{dx}{dt} + x = 0.$$

Express the equation as a two-dimensional continuous dynamical system. Analytically find the stability of the system's fixed point. Then show (again analytically) that it undergoes a Hopf bifurcation, and find at which value $\mu_c$ this happens. *Hint: what should happen to the Jacobian eigenvalues at a Hopf bifurcation?*

4.5 Consider the following system

$$\dot{x} = -x - y, \quad \dot{y} = -pz + ry + sz^2 - z^2y, \quad \dot{z} = -q(x + z)$$

introduced by Maasch and Saltzman [70] with default parameters $p = 1, q = 1.2, r = 0.8, s = 0.8$. Using Newton's method and the continuation described in Sect. 4.2.2 produce a bifurcation diagram for $x$ versus $p$ from 0 to 2. Because in this system it is possible to calculate the fixed points and their stability analytically, use this fact to confirm that your Newton's method for multidimensional systems is implemented correctly.

4.6 Explore a homoclinic bifurcation. Determine the fixed points of the system

$$\dot{x} = y, \quad \dot{y} = -py + x - x^2 + xy$$

and their stability as function of the parameter $p$. Compute the limit cycle occurring for $p = 1$ and its period. Then reduce the value of $p$ in small steps and monitor the change of the shape and the period of the limit cycle until a homoclinic bifurcation occurs. What happens to the period as we get closer and closer to the bifurcation point?

4.7 Compute and plot the basin of attraction of the Hénon chaotic attractor (for $a = 1.4, b = 0.3$). *Hint: safely assume that trajectories where the state vector $(x,y)$ obtains norm larger than 5 will escape to $\infty$.*

4.8 Continuing from above, find the eigenvectors of the unstable fixed point of the Hénon map for standard parameters. Evolve and plot its stable manifold and confirm that it forms the boundary of the basin of attraction of the chaotic attractor. *Hint: initialize several orbits along the direction of the stable eigenvector and evolve them backwards in time using the time-inverted Hénon map.*

4.9 Consider the function $h(z) = z^3 - 1$ defined on the complex plane, $z \in \mathbb{C}$. Use Newton's method to find the three zeros of $h$. Then, interpret the Newton method as a two-dimensional discrete dynamical system. Prove it is dissipative. Show that its fixed points coincide with the zeros of $h$. For these three fixed points find and plot in different color their basins of attraction on the square $[-1.5, 1.5]^2$. You should find that the basins of attraction have fractal boundaries, something we will discuss more in Sect. 5.3.3.

4.10 Program a function that produces the orbit diagram of the logistic map given the number $n$ of states to plot and $t$, a transient amount of steps to evolve the system for before plotting (start always from $x_0 = 0.4$). Keeping $n$ fixed, create an animation of how the orbit diagram changes when increasing $t$. What do you observe? For which values of $r$ is the convergence slowest?

4.11 Using the above orbit diagram, numerically estimate $\delta$ and $r_\infty$. Approximate infinity by some high period like $2^7$, while ensuring that your code runs with big floats (i.e., more than 64-bit floating numbers) for higher accuracy.

4.12 Produce orbit diagrams for the following maps: $x_{n+1} = r\sin(\pi x_n)$, $x_{n+1} = r\cos(\pi x_n)$, $x_{n+1} = x_n e^{-r(1-x_n)}$, $x_{n+1} = e^{-rx_n}$. Which ones show the period doubling route to chaos, which ones don't, and why?

4.13 Consider the Gissinger system [10]

$$\dot{x} = \mu x - yz, \quad \dot{y} = -\nu y + xz, \quad \dot{z} = \gamma - z + xy$$

(by default $\mu = 0.119$, $\nu = 0.1$, $\gamma = 0.9$ and initial condition $(3, 3, 3)$). Using the Poincaré surface of section produce an orbit diagram of the system for $\mu \in (0, 0.5)$ and $\nu \in (0, 0.5)$ (keep the other parameters at default values in each case), just like we did in Fig. 4.5 for the Rössler system.

4.14 Prove that when a period doubling bifurcation occurs in a map $x_{n+1} = f(x_n)$, the resulting two fixed points of $f^{(2)}$ form a period-2 cycle in $f$. Also prove that a fixed point of $f$ is a fixed point of $f^{(2)}$ with same stability properties. *Hint: chain rule for the win!*

4.15 In general, show that the existence of a fixed point of $f^{(n)}$ must necessarily mean the existence of an $n$-period in $f$ and that the stability of the periodic orbit is also the same as the stability of $f^{(n)}$ at the fixed point.

4.16 For $r$ close to $r_c = 1 + \sqrt{8}$ (but $r < r_c$) the logistic map displays intermittent behavior with the laminar phases being a period-3 orbit. Devise a way to segment the timeseries into laminar phases and chaotic bursts and estimate the mean laminar length $\langle \ell \rangle$. Repeat the process for various $r$ values. How does $\langle \ell \rangle$ depend on $\delta r = 1 + \sqrt{8} - r$?

4.17 Repeat the above exercise for the map of (4.5), versus parameter $\varepsilon > 0$.

4.18 Continuing from the above exercise, simulate timeseries of the map and then calculate their power spectrum. Fit a power law to the spectrum and estimate the power law exponent $\alpha$. How does $\alpha$ depend on $\varepsilon$? Similarly, compute the distribution of laminar lengths $\ell$, which you should also find a power law, $P(\ell) \sim \ell^{-\beta}$. Perform a fit and estimate how $\beta$ depends on $\varepsilon$.

# Chapter 5
# Entropy and Fractal Dimension

**Abstract** It seems reasonable to describe the chaotic and unpredictable nature of a dynamical system with probabilistic concepts. It turns out that this is indeed very fruitful, allowing one to quantify the complexity of the dynamics. Therefore, in this chapter we study dynamical systems through the lens of information theory. After introducing the concepts of information and entropy, we look into ways with which to probabilistically describe a deterministic dynamical system. Then, we expand on practically using these concepts to define dimensionality of chaotic sets, a concept that is taken further advantage of in following chapters. Because the structure of chaotic sets is intimately linked with fractal geometry, in this chapter we also introduce fractals.

## 5.1 Information and Entropy

### 5.1.1 Information Is Amount of Surprise

How can we quantify information content of an observation without referring to the subjective value it may have for some individual? The key idea is to quantify the amount of information $I$ obtained by the degree to which the outcome of an observation is surprising for the receiver (or the amount of uncertainty concerning the outcome of the observation). This concept of "surprisedness" may seem subjective, but it is in fact quantified by the probability $P$ of a particular observation (e.g., the outcome of a measurement or a computer simulation). If you receive always the same outcome (i.e., with probability $P = 1$) you don't get any new information about the source, and the information $I(P)$ provided by the observation is zero, i.e. $I(1) = 0$, because you are "zero surprised". On the other hand, if you investigate an unknown system and 9999 out of 10000 measurements tell you that the system is in some state A, but one outcome indicates that the system is in state B, then this outcome (which

© The Author(s), under exclusive license to Springer Nature Switzerland AG 2022

G. Datseris and U. Parlitz, *Nonlinear Dynamics*, Undergraduate Lecture Notes in Physics,

https://doi.org/10.1007/978-3-030-91032-7_5

occurs with very low probability) carries a particularly high information, because it is the only observation that tells you that state B is possible in the system of interest. It thus "surprises" you greatly.

Think about an ongoing global epidemic, and all inhabitants of your city are tested for the virus. The outcomes of these measurements are all negative (no infection), except for one poor soul who is positive. This outcome, which comes with a very low probability, is the outcome that matters and provides crucial information about the current state of your city (i.e., it is the only outcome that shows that the epidemic has reached your city).

## 5.1.2  Formal Definition of Information and Entropy

Considerations like the above mean that the information $I(P)$ has to be a positive, monotonously decreasing function of the probability $P$. And if two independent observations are made whose outcomes occur with probabilities $P_1$ and $P_2$ then the information $I(P_1 P_2)$ provided by both observations (where $P_1 P_2$ is the joint probability of independent processes) should be the sum of the information provided by the individual observations, $I(P_1) + I(P_2)$. These requirements, along with $I(1) = 0$ (absolute certainty has no information), are fulfilled by the relation

$$I(P) = -\log(P). \tag{5.1}$$

The base of the logarithm can be chosen arbitrarily and decides the measurement unit of $I$. For example, base 2 means units of "bits", so the information of an ideal coin toss with $P = 0.5$ is $I(0.5) = -\log_2(0.5) = 1$ bit.

Now we consider a distribution of different observations occurring with probabilities $P_m$, $m = 1, \ldots, M$. Here we assume that the number of possible observations $M$ is finite, see below for a justification. If many observations are made from this source then the average amount of information received from this source is given by the weighted average of $I$, called the *Shannon entropy*

$$H = -\sum_{m=1}^{M} P_m \log(P_m) \tag{5.2}$$

which vanishes if only one and the same observation labeled by index $n$ occurs all the time ($P_n = 1$, $P_{m \neq n} = 0$) and attains its maximum $H = \log(M)$ if all probabilities are the same, $P_m = 1/M$. $H$ may also be interpreted as a measure of disorder or "chaoticity" (see next section). Note that when practically computing the Shannon entropy (5.2) any term with $P_m = 0$ is omitted in the sum (use the Taylor expansion of the logarithm to justify this rule).

### 5.1.3 Generalized Entropy

The Shannon entropy defined in (5.2) can be extended to the so-called *generalized* or *Rényi entropy*

$$H_q(\varepsilon) = \frac{1}{1-q} \log \left( \sum_{m=1}^{M} P_m^q \right) \tag{5.3}$$

which for $q = 1$ reduces to (5.2). Varying $q$ puts more or less emphasis on high or low probabilities, and this feature can be used to characterize the structure of multi-fractal sets (see Sect. 5.4).

## 5.2 Entropy in the Context of Dynamical Systems

To study the entropy content of a set of points one must associate some probabilities $P_m$ to the data at hand, represented by a set $X = \{\mathbf{x}_1, \ldots, \mathbf{x}_N\}$, $\mathbf{x}_i \in \mathbb{R}^D$ which could, for example, be obtained from a trajectory of a dynamical system sampled at times $\{t_1, \ldots, t_N\}$. This is the case we want to consider in the following and there are several ways to create the probabilities $P_m$ (see below). Once we have them, we can calculate the entropy of $X$ using (5.3).

But here $X$ contains in fact real numbers, and thus observing a state $\mathbf{x}$ has infinite, and not finite, possible outcomes. Thankfully, for the purposes of this book this is not a big problem. Firstly, because all numbers in a computer are stored with finite precision, but most importantly, because in all the ways we will define $P_k$ for dynamical systems, we always constrain our numbers (or sequence of numbers) to a finite pool of possible observations context. Throughout the rest of this chapter it is best practice to standardize $X$. To do so, consider $X$ as a $(N, D)$ matrix and transform each column to have 0 mean and standard deviation of 1.

### 5.2.1 Amplitude Binning (Histogram)

The simplest way to obtain probabilities $P_m$ for $X$ is to partition the $D$-dimensional space into boxes of edge length $\varepsilon$ and then, for each box $m$, we count how many elements of $X$ are inside this box, $M_m$. To put it bluntly, we make a *histogram*[1] of $X$. For large $N$ the probability to find a point of $X$ within the $m$th box is approximately

$$P_m(\varepsilon) \approx M_m(\varepsilon)/N. \tag{5.4}$$

What we did above is sometimes referred to as "amplitude binning", or just binning, because we have binned the amplitudes (i.e., values) of the elements of $X$ based on rectangular boxes in the state space.

---

[1] In Chap. 8 we will discuss that this histogram of $X$ is approximating the natural density, and is connected with one of the fundamental properties of chaos, ergodicity.

In this context, the Shannon entropy $H_1(\varepsilon)$ of these $P_m$ quantifies the average amount of information we need to locate a point within $X$ with precision $\varepsilon$. The higher the entropy of $X$, the more random $X$ is, or the more complicated the patterns are the points in $X$ represent. The set of 2D points making a circle has lower entropy than a chaotic attractor of a 2D system, which in turn has less entropy than a set of 2D random numbers. In fact, this is a rather straightforward way of comparing the disorder of sets that have the same state space dimensionality and amount of points, and can even be used to quantify how the disorder of a dynamical system changes as a parameter changes.

To calculate the Rényi entropy (5.3), one only needs all probabilities $P_m$. As pointed out above, in principle this process amounts to making a histogram of the set $X$, by partitioning $\mathbb{R}^D$ into boxes, and then counting how many entries reside within each box. However, not all boxes that partition $\mathbb{R}^D$ are necessary for our goal here, as most of them will be empty, and thus can be skipped. This also avoids the typical memory overflow of naive histograms, whose memory allocation scales as $(1/\varepsilon)^D$. In Code 5.1 we present an algorithm to compute the probability $P_m$, which has memory allocation $O(DN)$ and performance $O(DN\log(N))$ and which is to our knowledge the fastest that exists.

**Code 5.1** Fast algorithm for the computation of Eq. (5.4).

```
using DynamicalSystems
function smarthist(X::Dataset, ε)
    mini, N = minima(X), length(X)
    P = Vector{Float64}()
    # Map each datapoint to its bin edge and sort them
    bins = map(point -> floor.(Int, (point - mini)/ε), X)
    sort!(bins, alg=QuickSort)
    # Fill the histogram: count consecutive equal bins
    prev_bin, count = bins[1], 0
    for bin in bins
        if bin == prev_bin
            count += 1
        else
            push!(P, count/N)
            prev_bin, count = bin, 1
        end
    end
    push!(P, count/N)
    return P
end
```

So how does Code 5.1 work? We first map all data points to the box they occupy. Typically many boxes are created more than once in this process, and to obtain $M_m$ we need to count how many instances of box $m$ we have created. The fastest way to do this is to first sort the boxes (by dimension: sort first by $x$, if $x$ is the same sort by

$y$, etc.) and then to count the subsequently equal boxes. Notice that Code 5.1 does not return the actual bin edges, as they are not needed. If needed, they are obtained by the command `unique!(bins) .* ε .+ mini`.

### 5.2.2 Nearest Neighbor Kernel Estimation

An alternative method to estimate probabilities $P$ used frequently in the literature is called nearest neighbor kernel estimation. The idea is to assign to each point $\mathbf{x}$ a probability $P$ based on how many other points are near it, i.e., for each $\mathbf{x}_i$ in the dataset define

$$P_i(\varepsilon) = \frac{1}{N - 2w - 1} \sum_{j:|j-i|>w} B\left(||\mathbf{x}_i - \mathbf{x}_j|| < \varepsilon\right) \tag{5.5}$$

where $|| \cdot ||$ represents a distance metric (typically Chebyshev or Euclidean) and $B = 1$ if its argument is true, 0 otherwise.[2] Here $w$ is called the *Theiler window*, illustrated in Fig. 5.1. It eliminates spurious correlations coming from dense time sampling of a continuous dynamical system and its value depends on how dense the sampling is. Typically a small integer is adequate, e.g., the first zero or minimum of the autocorrelation function of one of the timeseries of $X$. For discrete systems temporal neighbors are usually not neighbors in state space and you can choose $w = 0$. Just like the binning method, this method also depends on the parameter $\varepsilon$, the radius below which points are considered "neighbors". This parameter must be chosen at least as large as the minimum distance between any points in the dataset.

## 5.3 Fractal Sets in the State Space

"Dimension" is a word that can be used in many different ways (common downside of using human language) and one should take care to always specify it within context. For example, in Chap. 1 we used "dimension" simply as the number of independent variables of a dynamical system, i.e., the dimensionality of the state space $S$. However in the context of dynamical systems, dimension is also used to specify a dynamically invariant property of a set in the state space (typically the set a trajectory asymptotically covers during its evolution). This version of the dimension is called the *fractal dimension*, which is different from the topological dimension. The latter is the version that we are most familiar with and goes as follows: isolated points are zero-dimensional, lines and curves are one-dimensional, planes and surfaces are two-dimensional, etc. The topological dimension therefore (whose proper mathematical definition is not necessary for the current discussion) is always

---

[2] In the literature the Heaviside step function $\Theta$ with $\Theta(x) = 1$ for $x > 0$ and 0 otherwise is often used instead of $B$.

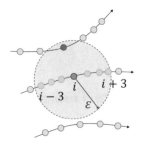

**Fig. 5.1** Illustration motivating the Theiler window. Around the black point, with index $i$, a disk of radius $\varepsilon$ is drawn. Orange and cyan points are within distance $\varepsilon$, however the cyan points should be excluded as they are neighbors of the black state $i$ only due to the fact they lie on the same trajectory segment. Such configurations occur typically around any state space point and wrongly bias calculations, because they make the number of neighboring points to scale like for a one dimension object (i.e., the set "looks" locally like a line). For the case shown in this sketch, at least $w = 3$ should be used. Black arrows are a guide to the eye, only the points (sampled trajectory) exist

an integer, corresponding to the number of independent variables or coordinates we need to describe a set.

### 5.3.1  Fractals and Fractal Dimension

The topological dimension does not provide a satisfactory description of the sets that arise when a chaotic dynamical system is evolved, because it does not take into account the complexity of the sets and thus cannot quantify it. What do we mean by "complexity"? Some sets have the property of having a *fine structure*, i.e., *detail on all scales*. This means that zooming into such set does not resolve its geometry into simpler patterns. Sets that satisfy this property are typically called *fractals*.

Let's make things concrete by comparing the plain circle to the *Koch snowflake*, which is a fractal set. We know how to make circles: they are specified by the geometric condition of all points with equal distance to a center. The Koch snowflake is specified differently: instead of having a geometric condition, we have an iterative process. We start with an equilateral triangle. Then, for each line of the shape, we replace the middle 1/3rd of the line with two lines of length 1/3rd of the original line, forming a smaller equilateral triangle. In the next "step" we repeat this exact same process for all lines that have resulted from the previous step. The process is illustrated in Fig. 5.2. The actual Koch snowflake is the set resulting after an infinite amount of these recursive steps. In the bottom row of Fig. 5.2 we zoom into the Koch snowflake and a circle. Fair enough, the circle eventually becomes a straight line. But this is not the case for the Koch snowflake. In fact, *by construction*, it doesn't matter how much we zoom, we will always see the same shape at exactly the same level of detail (or "complexity"). This feature is called *self-similarity*.

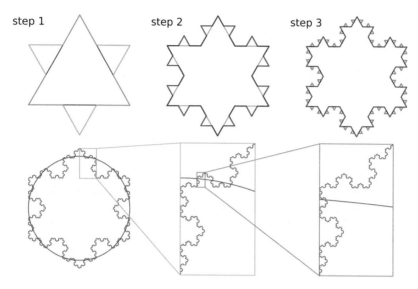

**Fig. 5.2** The Koch snowflake fractal. Top: construction process. Bottom: sequential zoom into the fractal and a circle

Now we can ask an interesting question: what is the dimension of the Koch snowflake? On one hand, the Koch snowflake is just a jagged line (even at the limit of infinite steps), so clearly its topological dimension is 1, as it can be described only by a single number, the length along the (jagged) line. It is however a line of infinite length, since at each step its length increases by a factor of 4/3. On the other hand, this infinitely long line is contained within a finite amount of space (as is evident by the figure) which in some sense would make it behave "like a surface". So it must have dimension more than one, but this contradicts the first statement...

This controversy is resolved by introducing a new kind of dimension, called the *fractal dimension*. For typical sets like cones, spheres and lines, this dimension coincides with the topological dimension of the set and is an integer. For irregular sets, this dimension is a non-integer number. Rigorous mathematical definitions for a fractal dimension can be given in terms of the Hausdorff dimension, but this goes beyond the scope of this book. Here we will use practical and computable definitions in Sect. 5.4. Using these definitions one finds for the Koch snowflake a fractal dimension of $\log(4)/\log(3) \approx 1.262$.

## 5.3.2 Chaotic Attractors and Self-similarity

Dissipative systems that display deterministic chaos have attractors that are chaotic. These objects are often fractals and are called *chaotic* or *strange attractors*, although other "non-strange" and "non-chaotic" variants also exist. The Hénon attractor in

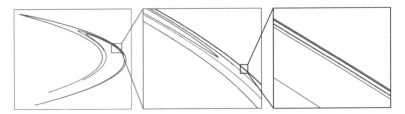

**Fig. 5.3** Self-similarity of the Hénon attractor, (3.3). The construction of the Hénon map, shown in Fig. 3.6 explains the origin of the self-similarity

Fig. 5.3 is a chaotic attractor. For this set, it is straightforward to see self-similarity by zooming in the set. The self-similarity here stems from the stretching and folding mechanisms of chaos, Sect. 3.2.4. Of course, thematic connection to self-similarity also exists at a fundamental level because dynamical systems are also iterative processes: the current state defines the next one, and the next one defines the second-next one using the same rules.

Notice the difference between "perfect" self-similarity, displayed by the Koch snowflake, and the qualitative one of the Hénon map attractor. Fractal sets arising from dynamical systems are never exactly the same under zooming (simply because the rule $f$ depends on the state space location $\mathbf{x}$, and thus different parts of the state space are transformed differently).

### 5.3.3 Fractal Basin Boundaries

In Chap. 1 we defined the basin of attraction (of an attractor) as the set of all points that converge to the given attractor.[3] In Fig. 5.4, we show the basins of attraction of the magnetic pendulum, a system with a magnet on a string hovering above $N = 3$ fixed magnets on a plane at locations $(x_i, y_i)$. The dynamic rule is

$$\dot{x} = v_x, \quad \dot{v}_x = -\omega^2 x - q v_x - \sum_{i=1}^{N} \frac{\gamma (x - x_i)}{d_i^3}$$

$$\dot{y} = v_y, \quad \dot{v}_y = -\omega^2 y - q v_y - \sum_{i=1}^{N} \frac{\gamma (y - y_i)}{d_i^3}$$

with $d_i = \sqrt{(x - x_i)^2 + (y - y_i)^2 + d^2}$ (we used $q = 0.2, \omega = 1.0, d = 0.3, \gamma = 1$). Because of the friction term $qv$, the pendulum will stop swinging at some point.

---

[3] A straightforward way to calculate basins of attraction is as follows: create a grid of initial conditions that is a sufficiently dense covering of the state space. Then evolve each initial condition for long times, recording the attractor that it "settles in" at the end. All points that settle in the same attractor belong to its attraction basin.

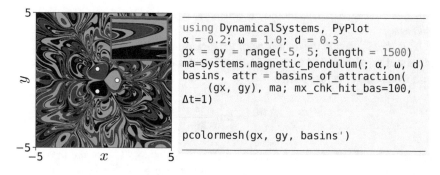

```
using DynamicalSystems, PyPlot
α = 0.2; ω = 1.0; d = 0.3
gx = gy = range(-5, 5; length = 1500)
ma=Systems.magnetic_pendulum(; α, ω, d)
basins, attr = basins_of_attraction(
    (gx, gy), ma; mx_chk_hit_bas=100,
Δt=1)


pcolormesh(gx, gy, basins')
```

**Fig. 5.4** Basins of attraction of the three magnets (white dots) of the magnetic pendulum. Each point of the $(x, y)$ plane corresponds to an initial condition $(x, y, 0, 0)$. The point is colored according to which of the three magnets the orbit starting at this initial condition converges to. An inset displays an 80-fold zoom of the plot. Important clarification: it is not the colored regions that constitute the fractal object, it is their *boundaries*

And because of the magnetic field of the magnets, the pendulum will stop above one of the three magnets.

Thus, in the long term, the system is not chaotic, as it will stabilize in a fixed point. But which point, that is very tough to say, as we can see in Fig. 5.4. The magnetic pendulum is an example of a system with *multi-stability* (many stable states for a given parameter combination) but also an example of a system that displays *transient chaos*. This means that while the long term evolution of the system is not chaotic, there is at least a finite time interval where the motion displays sensitive dependence on initial conditions. In the current example, this time interval is the start of evolution because all three stable fixed points have *fractal boundaries* for their basins of attraction. A consequence is that the magnetic pendulum has *final state sensitivity*, that is, what the final state of the system will be depends sensitively on the initial condition.

## 5.4 Estimating the Fractal Dimension

Practical estimates of the fractal dimension will be denoted by $\Delta$. But how does one even start to create a definition of a fractal dimension? We can be inspired by one of the features the "classic" version of a dimension quantifies, which is the scaling of a volume of an object if we scale it all up by some factor. Scaling a line by a factor of 2 increases its "volume" (which is its length) by 2. Doing the same to a square increases its "volume" (which is its area) by a factor $2^2$, while for a cube this increase is $2^3$. The dimension is the exponent. The simplest way to estimate this exponent is to cover the set with non-overlapping (hyper-)boxes of size $\varepsilon$ and count the number of boxes $M$ we need to cover the set. Then $M$ generally scales as $M \propto (1/\varepsilon)^\Delta$, as is illustrated in Fig. 5.5. This method is called the *box-counting dimension* or *capacity dimension*.

This idea is used in the definition of the *generalized dimension* $\Delta_q^{(H)}$. It is defined as the scaling of the Rényi entropy $H_q$, calculated with the binning method of Sect. 5.2 with a box size $\varepsilon$, versus $\log(\varepsilon)$

$$\Delta_q^{(H)} = \lim_{\varepsilon \to 0} \frac{-H_q(\varepsilon)}{\log(\varepsilon)}. \tag{5.6}$$

The capacity or box-counting dimension ($q = 0$, $H_0 = \log(M)$) or the *information dimension* ($q = 1$) are commonly used in the literature. $\Delta_q^{(H)}$ is a non-increasing function of $q$ and the set is called *multi-fractal* if $\Delta_q^{(H)}$ changes with $q$. Most chaotic sets are multi-fractal.

A different approach to define a fractal dimension is the *correlation dimension* $\Delta^{(C)}$, and is based on the *correlation sum* (related to (5.5)). Similarly with $\Delta_q^{(H)}$ we first calculate the correlation sum

$$C(\varepsilon) = \frac{2}{(N-w)(N-w-1)} \sum_{i=1}^{N-w-1} \sum_{j=1+w+i}^{N} B(\|\mathbf{x}_i - \mathbf{x}_j\| < \varepsilon) \tag{5.7}$$

with $B = 1$ if its argument is true, 0 otherwise. Notice that $C$ is in fact just the average of (5.5) for all points in the dataset. The correlation dimension $\Delta^{(C)}$ is then defined by a similar scaling law as (5.6), $\Delta^{(C)} = \lim_{\varepsilon \to 0} (\log(C)/\log(\varepsilon))$. In Fig. 5.6 we show how $H_2$ and $C$ behave for different $\varepsilon$ for the attractor of the Hénon map.

Since all datasets are finite, the theoretical limit of $\varepsilon \to 0$ cannot be achieved, as there exists some $\delta = \min_{i \neq j}(\|\mathbf{x}_i - \mathbf{x}_j\|)$ below which it is meaningless to compute $H_q$ or $C$, because $H_q$ reaches its maximum value, while $C$ becomes 0. This is seen in the left part of Fig. 5.6, where the $H_2$ curve flattens for small $\varepsilon$ (while for even smaller $\varepsilon$ the correlation sum $C$ would become 0, and $\log(C) \to -\infty$). $\varepsilon$ also has an upper limit: if it is larger than the extent of the dataset itself, it again makes no

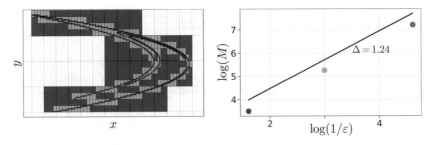

**Fig. 5.5** Covering the Hénon attractor with squares of different size $\varepsilon$, and counting the number $N$ necessary to cover the whole attractor. The scaling of $M$ versus $\varepsilon$ defines the capacity or box-counting dimension (left plot is not in equal aspect ratio)

sense to compute $H_q$, as it will always be zero (in (5.3) we get log(1)), while $C$ will always be 1. Thus one must choose an appropriate range of $\varepsilon$ that indicates a *linear scaling region* for $H_q$ versus $\log(\varepsilon)$, as shown in Fig. 5.6. See also Sect. 12.3.2, which discusses how difficult it is to estimate fractal dimensions for real data.

An alternative method for computing $\Delta$ is *Takens' best estimator*

$$\Delta^{(T)} \approx -\frac{1}{\eta}, \quad \eta = \frac{1}{N^*} \sum_{[i,j]^*} \log \left( \frac{||\mathbf{x}_i - \mathbf{x}_j||}{\varepsilon_{\max}} \right) \tag{5.8}$$

where the summation happens for all $N^*$ pairs $i$, $j$ such that $i < j$ and $||\mathbf{x}_i - \mathbf{x}_j|| < \varepsilon_{\max}$. It exchanges the challenge of having to find an appropriate scaling region in $\varepsilon$ with the task to identify a threshold $\varepsilon_{\max}$ (where linear scaling of $C$ stops). Knowing $\varepsilon_{\max}$ a priori in the general case isn't possible.

## 5.4.1 Why Care About the Fractal Dimension?

There are several reasons! The most important one is that if the dataset $X$ is obtained from an unknown system through some measurement, then the ceiling of $\Delta$ (the next integer larger than $\Delta$) represents a strict lower limit on the amount of independent variables that compose the unknown system. This can be a guiding factor in any modelling process. Equally importantly, $\Delta$ approximates the effective degrees of freedom of the system. For example, the Lorenz-63 chaotic attractor of Fig. 1.1 has $\Delta \approx 2$. This means that, while living in 3D space, the chaotic attractor is locally very flat, like a plane. Same is true for the Rössler chaotic attractor of Fig. 4.7. $\Delta$ is also a dynamic invariant, like the Lyapunov exponents, i.e., it retains its value if we transform the set via diffeomorphisms, while standard statistical quantifiers like means or standard deviations would change. Furthermore, the method of estimating $\Delta$ can be used to estimate the noise level of data that are assumed to be deterministic

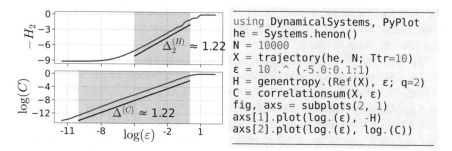

**Fig. 5.6** Scaling of $H_2$ and $\log(C)$ versus $\varepsilon$ for the Hénon attractor (3.3) with a light orange background indicating the linear scaling regions. A slope of 1.22 fits both curves (red line)

with added noise (see below). Lastly, $\Delta_0^{(H)}$ is theoretically tied with the foundation of nonlinear timeseries analysis, see Chap. 6.

### 5.4.2  Practical Remarks on Estimating the Dimension

Before even starting to compute a fractal dimension, one must make sure that the set at hand describes a *stationary* system (i.e., where the dynamical rule does not change over time). This is critical, because the fractal dimension is quantifying an invariant set in the state space, and we assume that our measurements are sampling this set, and not that each measurement samples a different version of the set. Furthermore, if the dynamics is not stationary, then the probabilities discussed in Sect. 5.2.1 are ill-defined, as they would change in time, while we assume that the $P_m$ are constants that we estimate better and better with increasing data length $N$.

Regarding $N$, the minimal amount of points $N_{min}$ required to accurately estimate a dimension $\Delta^{(C)}$ roughly depends exponentially on the dimension, $\log(N_{min}) \propto \Delta^{(C)}$. This means that to have a statistically meaningful scaling law of $C$ covering at least an order of magnitude of $\varepsilon$ in base 10, an amount of points of roughly $N_{min} \sim 10^{\Delta}$ is necessary.[4] This has the implication that one may underestimate the dimension of a set due to lack of data.

### 5.4.3  Impact of Noise

When noise comes into place, there is one more thing to take into account. In general, the fractal dimension depends on, and quantifies, the fine structure of a given set. However, the smallest scales of a noisy set are dominated by the structure of the noise, and not of the (potentially) fractal structure of the set. This means that the slope of the correlation sum at the smallest $\varepsilon$ is *not* the fractal dimension of the dynamical system's set, but of the noise, whose dimension typically coincides with the state space dimension. And noise doesn't have to be only observational noise, from measurements of real world systems, because low algorithmic accuracy can also produce noise in data.

We illustrate this issue in Fig. 5.7, which shows the Poincaré section of the Rössler system computed with different algorithmic accuracy. Given the discussion of Sect. 4.3.1, the strange attractor of the Rössler system has a fractal dimension $\approx 2$. As a result, its Poincaré section must have a dimension of $\approx 2 - 1 = 1$. In Fig. 5.7 we see that for the noiseless estimate, we find a correct value, while for the noisy one the small scales have slope much closer to 2 (which is the expected fractal dimension of noise in 2D), and only at larger $\varepsilon$ the slope comes closer to 1. This allows one

---

[4] As you will see in the exercises, this strict condition is a sufficient, not a necessary condition. For many datasets you may get quite accurate estimates of $\Delta$ with much less points.

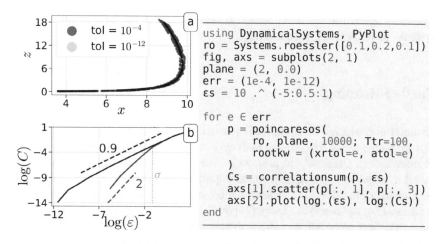

```julia
using DynamicalSystems, PyPlot
ro = Systems.roessler([0.1,0.2,0.1])
fig, axs = subplots(2, 1)
plane = (2, 0.0)
err = (1e-4, 1e-12)
εs = 10 .^ (-5:0.5:1)

for e ∈ err
    p = poincaresos(
        ro, plane, 10000; Ttr=100,
        rootkw = (xrtol=e, atol=e)
    )
    Cs = correlationsum(p, εs)
    axs[1].scatter(p[:, 1], p[:, 3])
    axs[2].plot(log.(εs), log.(Cs))
end
```

**Fig. 5.7  a** Poincaré section of the Rössler system, exactly like in Fig. 4.7, for two different interpolation tolerances (for finding the actual section) given in the function `poincaresos` . **b** Estimates of the fractal dimension of **a** using the correlation sum, with $\sigma$ denoting the observed noise radius

to identify a *noise radius*, i.e., how large the scale of the noise is, which is when the scaling changes from that of the noise to that of the fractal set, as illustrated in Fig. 5.7 with $\sigma$. This is useful even in real world applications where one does not know the "expected" fractal dimension a-priori, because it is always the case that noise will have higher fractal dimension (and thus the point of change of slopes can still be identified). An important tool for distinguishing deterministic from stochastic components are surrogate data (tests) that will be introduced in Sect. 7.4.

### 5.4.4  Lyapunov (Kaplan–Yorke) Dimension

An alternative dimension definition, that doesn't come from the state space structure of the set, but from the chaotic dynamics, is known as *Lyapunov dimension* or *Kaplan–Yorke dimension*. It is defined as the point where the cumulative sum of the Lyapunov exponents of the system first becomes negative, and thus only applies to dissipative systems. This point can be found via linear interpolation by

$$\Delta^{(L)} = k + \frac{\sum_{i=1}^{k} \lambda_i}{|\lambda_{k+1}|}, \quad k = \max_{j} \left[ \sum_{i=1}^{j} \lambda_i > 0 \right]. \tag{5.9}$$

The good thing about (5.9) is that it remains accurate and straightforward to compute even for high-dimensional dynamics, provided that one knows the dynamic rule $f$ (because only then accurate algorithms exist for calculating the entire

Lyapunov spectrum, see Sect. 3.2.2). This is especially useful for spatiotemporal systems, as we will see in Sect. 11.3.1.

## Further Reading

Shannon is considered the founder of modern information theory [71, 72]. He was the first to discuss these concepts on mathematical footing and to define $I = -\log(p)$ as information. Originally in this work "outcome of an observation" was instead called a "message". Rényi also made important contributions and introduced the generalized dimension [73]. A textbook oriented more towards information theory is *Elements of Information Theory* [74] by Cover and Thomas. Many more methods exist to define probabilities from a dynamical system, see the documentation of DynamicalSystems.jl for numeric implementations, and the review on the "Entropy universe" by Ribeiro et al. [75].

The following three textbooks are recommended for reading more about the fractal dimension. The first is *Nonlinear Time Series Analysis* [76], which discusses estimating fractal dimensions with heavy focus on real-world data. The second is *Chaotic Dynamics: An Introduction Based on Classical Mechanics* [77], which discusses fractal sets arising from both dissipative and conservative systems, with focus on theory, multi-fractality, and the connection of the fractal set with its invariant density, which we will discuss later in Chap. 8. Last is *Deterministic Chaos: an introduction* by Schuster and Just [69], which discusses dynamical systems theory, and in particular dimensions and entropies in more mathematical details. There you can also find the definition of the Kolmogorov–Sinai entropy and the correlation entropy (which is a lower bound of the Kolmogorov–Sinai entropy).

To learn more about fractals, *Fractal Geometry Mathematical Foundations and Applications* by Falconer [78] is a good choice. Even though it is a mathematics textbook, it has several chapters on dynamical systems applications. Fractals were initially introduced by Mandelbrot in a highly influential work [79]. The concept of a non-integer dimension was given a rigorous mathematical footing by Hausdorff much earlier (the so-called Hausdorff dimension) [80]. We mentioned that chaotic attractors (typically called strange attractors) are fractal sets. Anishchenko and Strelkova [81] is a great discussion on the different types of attractors and how chaotic they are based on a sensible classification scheme (related to hyperbolicity, which we do not discuss in this book).

An exceptional review of the fractal dimension is given by Theiler in [82], which we highly recommend. There the different definitions of dimensions are connected with each other, as well as with the natural density (or measure) of chaotic sets, which will be discussed in Chap. 8. An up-to-date comparison of all existing methods to compute a fractal dimension, and their limitations and pitfalls, is given in [83]. The earliest report of a computationally feasible version of a fractal dimension (which nowadays is called the box counting dimension) was given by Pontrjagin and Schnirelmann in 1932 [84] (see [85] for more). The first version of the information dimension was given by Balatoni and Rényi in 1956 [73], in connection with the Shannon entropy. Application of fractal dimensions into dynamical systems context however started and flourished in the decade of 1980–1990s [86–92] with leading figures on the subject being Eckmann, Farmer, Grebogi, Grassberger, Ott, Procaccia, Ruelle, Takens, among many others (see [82] for more historic references and also the more recent Scholarpedia entry [93]). Eckmann and Ruelle [94] have shown that the number of points required to get an accurate estimate the fractal dimension depends exponentially on the fractal dimension itself (placing limitations of what dimensions you can estimate given a specific amount of points).

The correlation dimension $\Delta_2^{(C)}$ was originally introduced in the pioneering work of Grassberger and Procaccia [95] and shortly after improved by Theiler [96] and thus is often referred to as Grassberger–Procaccia dimension. Regarding the Lyapunov dimension, it was proposed by Kaplan and Yorke [97, 98] (hence it is often called "Kaplan–Yorke dimension"). They conjectured the equivalence between $\Delta^{(L)}$ and $\Delta_1^{(H)}$ (the information dimension).

Grebogi et al. [99] have devised a measure to examine the fractality of basin boundaries called *uncertainty exponent*, which is implemented as `uncertainty_exponent` in DynamicalSystems.jl. See also [100] for more info. The impact of noise on estimating the fractal dimension can be taken advantage of to estimate the signal-to-noise ratio as we have already shown, and see [76, Chap. 6] for further discussion.

If the fractal dimension depends on $q$, the set is called multi-fractal. This is the case when the natural measure underlying the attractor is non-uniform [91]. Large positive values of $q$ then put more emphasis on regions of the attractor with higher natural measure. The dependence of $\Delta_0^{(H)}$ on $q$ is connected with another concept, the so-called singularity spectrum, or multi-fractality spectrum $f(\alpha)$, which we did not discuss here, but see, e.g., Chap. 9 of [28] by Ott for more. A good reference for multi-fractality is the book by Rosenberg [101].

## Selected Exercises

5.1 Show that in (5.2) any term with $P_m = 0$ is omitted in the sum. Then show that the Rényi entropy (5.3) reduces to the Shannon entropy for $q \to 1$.

5.2 What is the Shannon entropy of a periodic orbit of a discrete dynamical system with period $m$?

5.3 Simulate the logistic map $x_{n+1} = 4x_n(1 - x_n)$ and the tent map

$$x_{n+1} = 2x_n, \text{ if } x_n \leq 0.5, \quad 2 - 2x_n \text{ else.}$$

and calculate and compare their histograms. Based on that figure, which system you expect to have higher Shannon entropy for a given $\varepsilon$? Confirm your findings numerically.

5.4 Show that the logistic map can be transformed to the tent map using the change of coordinates $T(x) = (1 - \cos(\pi x))/2$. Then numerically confirm that the histogram of the logistic map is transformed to that of the tent map, which you have computed in the previous exercise. What can you deduce about the dynamic invariance of the Rényi entropy from this?

5.5 A simple way to assign an entropy value to a timeseries is via the histogram as discussed in Sect. 5.2.1. Another simple alternative is to use the Fourier spectrum of the timeseries to quantify the complexity of its temporal variability. Transform a timeseries $x$ of length $N$ into a vector of probabilities by normalizing its power spectrum to have sum 1. Apply this process for a periodic timeseries, a chaotic timeseries and pure noise and then calculate their spectral entropy using these probabilities. In which case do you expect higher entropy? Do your numeric results confirm your idea? For a more advanced method that achieves a similar result see Sect. 6.4.3.

5.6 Using function recursion, write a code that generates arbitrary iterates of the Koch snowflake (by generating the vertices of the curve). For a high iterate, numerically calculate its capacity dimension $\Delta_0^{(H)}$, which you should find to be $\approx \ln 4/ \ln 3$.

5.7 Produce a figure similar to Fig. 5.6 for all three systems shown in Fig. 1.1. What is the fractal dimension you estimate for each set?

5.8 Continuing from the previous exercise, now focus on the Lorenz-63 system (1.5). At which size $\varepsilon_c$ does the linear scaling of the correlation sum stop? Quantify how does $\varepsilon_c$ depend on how large dataset you simulate initially, and/or the sampling time. *Hint: make a plot of $\varepsilon_c$ versus data length $N$, which you should find to be a power law.*

5.9 For the Lorenz-63 system you should have obtained a dimension value of $\Delta \approx 2$. Obtain a Poincaré surface of section (see Chap. 1) and calculate the fractal dimension of the section. Does the result give $\Delta - 1$?

5.10 The Lorenz-96 system [102] has a tunable state space dimension $D$ and is given by

$$\frac{dx_i}{dt} = (x_{i+1} - x_{i-2})x_{i-1} - x_i + F \qquad (5.10)$$

for $i = 1, \ldots, D$ and $D \geq 4$. The $D$th coordinate is connected with the first as in a loop. In your simulations use $F = 24$. For this parameter the system ends up in a chaotic attractor for almost all $D$ and initial conditions. Perform simulations for various $D$ and calculate the fractal dimension. Show that $\Delta$ increases approximately linearly with $D$ and find the proportionality coefficient.

5.11 Load exercise datasets 2, 3, 4, 6, 8, 9. Estimate their fractal dimension (and keep in mind all practical considerations!) using the correlation dimension.

5.12 Continue Exercise 3.5, and now also plot the Lyapunov dimension in addition to the Lyapunov exponents, versus the parameter. Identify parameter values with significantly different dimension, and deduce what the trajectories in state space should look like given the dimension value. Plot the trajectories for these parameters to confirm your suspicions.

5.13  Characterize the sensitivity of fractal dimension estimators on the orbit length
and initial conditions. Generate 5 *chaotic* initial conditions for the Hénon–
Heiles system, (1.6), ensuring they all have the same energy (find the Hamiltonian of the Hénon–Heiles system). In principle, all initial conditions belong to
the same chaotic set. Evolve each one for a total time of 10000 time units, with
$dt = 0.1$. Split each trajectory into segments of 1000 time units and calculate
the correlation dimension. Do the same for the full trajectory (so you should
have 60 values in total). How do the results change? Repeat the process (same
lengths and initial conditions) for the Lyapunov dimension. Is one method more
accurate than the other?

# Chapter 6
# Delay Coordinates

**Abstract** While models are great, and easy to tackle from many different directions, the real world is not so kind. When describing and measuring real systems, one often has access only to a single timeseries that characterizes the system. Fortunately for us, in most cases even from a single timeseries one can squeeze out knowledge about the original system, in a way reconstructing the original dynamical properties. In this chapter we discuss the way to do this, using delay coordinate embeddings. We particularly focus on successfully applying this tool to real world data, discussing common pitfalls and some nonlinear timeseries analysis techniques based on delay embedding.

## 6.1 Getting More Out of a Timeseries

In the dynamical systems that we introduced in Chap. 1, all variables were known at any time by (numerically) evolving an initial condition in the state space according to the rule $f$. In reality though, e.g., experimental measurements of real systems, one usually does not have access to all $D$ variables of the original state space $S$. Often one obtains only a timeseries of an observation $w$ sampled at times $t_n$, and even worse this $w$ might not even be one of the variables of the original system, but a function of the state. This is formalized by a measurement function $h : S \rightarrow \mathbb{R}$ which takes as an input the state $\mathbf{x}(t)$ of the original system, and outputs the observation $w$, i.e., $w(t) = h(\mathbf{x}(t))$.

Why is this a problem? Well, because quantities of interest (e.g., Lyapunov exponents, fractal dimensions) are defined in the state space and it seems that by reducing a high dimensional system to a single timeseries, which could be an almost arbitrary function of the true state of the system, we have lost all information about the full state space. It is definitely impossible to make any claims about the original dynamics with only $w(t)$... Just kidding! With what seems like pure magic we are in fact able to

G. Datseris and U. Parlitz, *Nonlinear Dynamics*, Undergraduate Lecture Notes in Physics,
https://doi.org/10.1007/978-3-030-91032-7_6

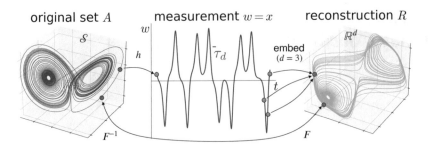

**Fig. 6.1** Delay coordinates embedding of a timeseries generated by the Lorenz-63 system (1.5) with delay time $\tau_d = \tau \delta t$ (red bar in middle panel shows the relative size of $\tau_d$) and $d = 3$. The figure illustrates how three points of the timeseries are combined to give a three dimensional point in the reconstruction space. An animation of this process can be seen online in animations/6/embedding

recover most dynamical properties of the original system. This magic (it is just math actually) is now known as *delay coordinates embedding* or just *delay coordinates*.

### 6.1.1   Delay Coordinates Embedding

What we'd like to do is to generate a new state space representation from the available timeseries. This is possible to do, because in deterministic dynamical systems, the past determines the future. This means that by looking at samples $w(t)$ along with their near future values $w(t + \Delta t)$, this should in some way yield information about how $f$ generates the original dynamics in $S$ and how it results in the set of states $A$ that $w$ represents. Let's put this into equations: from the samples of the timeseries $w$ (which has length $N$ and for continuous systems sampling time $\delta t$, $w[n] = w(n\delta t)$) we generate *delay vectors* $\mathbf{v}_n$ by combining the sample $w[n]$ with some more ones from the future (or past)

$$\mathbf{v}_n = (w[n], w[n + \tau], w[n + 2\tau], \ldots, w[n + (d - 1)\tau]) = \mathcal{D}_d^\tau w[n] \qquad (6.1)$$

where the dimension $d$ of $\mathbf{v}_n$ is called the *embedding dimension*, $\tau$ is the *delay time* (also called *lag*) and $\mathcal{D}_d^\tau$ stands for transformation of samples of a timeseries $w$ to a vector $\mathbf{v}$.

Equation (6.1) gives a new state space vector $\mathbf{v}_n$ from the timeseries, and its coordinates can be thought of as a new coordinate system to describe the dynamics (we will explain in the next section why this is valid to do). This process of state space reconstruction is illustrated in Fig. 6.1.

Note that with the notation used in (6.1) the delay time $\tau$ is an integer that gives for discrete systems the number of time steps or iterations between elements of $\mathbf{v}_n$ and for continuous systems the delay time in units of the sampling time $\delta t$ (the delay time in units of the system time $t$ is in this case $\tau_d = \tau \delta t$). Depending on the sign of

$\tau$ we can have either a forward $(+)$ or backward $(-)$ embedding. For the discussion of the present chapter, forward or backward is only a matter of taste, but in other cases it makes a difference (Sect. 7.2). In DynamicalSystems.jl a delay embedding of $w$ with $d$, $\tau$ is done by simply calling `embed(w, d, τ)`.

Now you may ask "Which values of the embedding dimension $d$ and the delay time $\tau$ (or $\tau_d = \tau \delta t$) shall I use?". The good news is that for almost all choices of measurement functions $h$ and times $\tau$ delay embedding provides a faithful and useful reconstruction of states, trajectories, and attractors, provided a sufficiently large embedding dimension $d$ is used. Here high enough $d$ is required to make sure that the dynamics properly unfolds in the reconstruction space without self-crossing of trajectories, for example. Which conditions for $d$ have to be fulfilled will be discussed in the next section and methods for choosing "optimal" values for $\tau$ and $d$ are introduced in Sect. 6.2.

### 6.1.2 Theory of State Space Reconstruction

To discuss the theoretical background of delay embedding we refer in this section to the delay time $\tau_d = \tau \delta t$ in time units of the dynamical system generating the observable $w(t) = h(\mathbf{x}(t))$. Using the flow $\mathbf{x}(t) = \Phi^t(\mathbf{x})$ with $\mathbf{x} = \mathbf{x}(0)$ the delay vector $\mathbf{v}(t)$ can be written as a function $F$ of the state $\mathbf{x}(t)$

$$\mathbf{v}(t) = (h(\mathbf{x}(t)), h(\Phi^{\tau_d}(\mathbf{x}(t))), \ldots, h(\Phi^{(d-1)\tau_d}(\mathbf{x}(t)))) = F(\mathbf{x}(t)) \in \mathbb{R}^d. \quad (6.2)$$

This function $F : A \to \mathbb{R}^d$ mapping states $\mathbf{x} \in A \subset S$ to reconstructed states $\mathbf{v} \in \mathbb{R}^d$ is called the *delay coordinates map* and its features are crucial for the concept of delay coordinates. For example, let's say you want to estimate the fractal dimension $\Delta$ of some (chaotic) attractor $A$ in the original state space. Then you may think of estimating $\Delta$ for the corresponding set $R = F(A) \subset \mathbb{R}^d$ of reconstructed states in $\mathbb{R}^d$ and ask under which conditions the value obtained for $R$ would coincide with the true but unknown fractal dimension of $A$. This is the case if the delay coordinates map $F$ is a *diffeomorphism* between $A$ and $R$, since the fractal dimension is invariant with respect to diffeomorphic transformations. The same is true for stability features and Lyapunov exponents. Now the question is, for which choices of $h$, $\tau_d$ and $d$ can we expect $F$ to be a diffeomorphism?

Fortunately, there are several *delay embedding theorems* answering this question. The good news is that, provided that you choose a sufficiently high embedding dimension $d$, then almost all smooth $h$ and $\tau_d$ will yield a diffeomorphic transformation. Lower bounds for suitable values of $d$ depend on the task or application. If, for example, the time series stems from a trajectory running on a chaotic attractor (or some other invariant set) $A$ then choosing $d > 2\Delta_0^{(H)}(A)$ is enough. This implies a one-to-one mapping of states excluding self-crossing of reconstructed trajectories which is important when using delay embedding for timeseries prediction (more on that in Sect. 6.4.1). Here $\Delta_0^{(H)}(A)$ is the capacity dimension of the original set, a concept that we defined and discussed in detail in Chap. 5. In contrast, if you are interested in only calculating the fractal dimension of the original set $A$, then the condition $d > \Delta_0^{(H)}(A)$ turns out to be sufficient. Notice that if the recorded time-series is coming from transient dynamics instead of an invariant set, you can still reconstruct the original dynamics. However in this case you may need to consider the full state space and use $d > 2D$.

Keep in mind that these conditions for the embedding dimension are *sufficient* conditions, not *necessary* ones. This means that if you satisfy these conditions, you are sure to be on the safe side. However, smaller embedding dimensions below these bounds may still be fine (see Sect. 6.2.2 for more).

The point we still have to discuss is the meaning of "almost all". Although this is mathematically more subtle it basically means that any pair $(h, \tau_d)$ picked "at random" is a good one for the purpose of embedding. But of course, this doesn't exclude bad examples such as embedding any coordinate of a $T$-periodic orbit $\mathbf{x}(t)$ with delay time $\tau_d = T$. In this case the reconstructed set would in fact be just the diagonal line, i.e., the reconstruction is not one-to-one.

In the same sense, there are observation functions $h$ that cannot yield a diffeomorphism for any $(d, \tau_d)$. An example is when $h$ collapses the set original $A$ under an existing symmetry which results in irrecoverable information loss. To illustrate, let's have a look at the chaotic Lorenz-63 attractor in Fig. 1.1. If our measurement function is $h(\mathbf{x}(t)) = z(t)$, then the two "wings" of the attractor are collapsed into one, because it is impossible to tell whether a given value of the $z$ coordinate is coming from either the left or right wing. Said differently, the dynamics of the Lorenz-63 system is symmetric under the transformation $(x, y, z) \mapsto (-x, -y, z)$, which exchanges the two wings with each other, but leaves $z$ invariant. Therefore this information, that in the original set two wings exist, cannot be recovered via delay coordinates embedding of $z$.

## 6.2  Finding Optimal Delay Reconstruction Parameters

We are faced with a paradox, which hopefully has been bugging you already. To find an appropriate $d$ to embed the timeseries, we need to know the fractal dimension of the attractor that generated said timeseries or the dimension of the state space of the observed system. But for an unknown (experimental) system in general we don't know neither of these dimensions! And to determine, for example, the fractal dimension, we need to already have a properly embedded set... So how can we ever reliably embed a timeseries?

The delay embedding theorems place very loose limitations, requiring us to only choose a sufficiently large embedding dimension $d$ and then we are "theoretically" in a fine scenario. They provide no information on how to actually choose the reconstruction parameters $d$ and $\tau$ though,[1] and in practice one has to find not only sufficient, but also optimal values of $d, \tau$. Here we will discuss the conventional approach to optimal delay embedding, which consists of first optimizing the choice of a (constant) delay time $\tau$ and then of an embedding dimension $d$ based on the found $\tau$. In Sect. 6.3.3 you learn about more advanced methods aiming at simultaneously optimizing the embedding parameters.

Notice that you have to make sure that the dynamics underlying your timeseries is *stationary*, which means that the underlying dynamical rules do not change with time. In addition, it is crucial that your timeseries is sampled with a *constant sampling rate*. In the discussion we had so far, $\tau_d$ is a constant, not a function of time! For delay embedding to be valid, the samples $(w[n + \tau], w[n])$ and $(w[m + \tau], w[m])$ must have exactly the same temporal difference $\tau_d = \tau \delta t$ for any $m, n$.

### 6.2.1 Choosing the Delay Time

Why do we need an optimal $\tau$? Let's imagine for a second that we have a fixed $d > 2\Delta_0^{(H)}$ and we vary $\tau$; in theory almost any $\tau$ value is sufficient. What ruins the fun in reality is that one has only a finite number of samples and thus reconstructed states. In addition, noise and discretization errors (finite-precision representation of real numbers and finite-time sampling of continuous systems) make matters even worse.

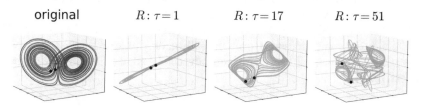

**Fig. 6.2** Impact of delay time on the same reconstruction process as in Fig. 6.1, with $\tau = 17$ the being optimal delay time (minimum of self-mutual information) and $\delta t = 0.01$. An interactive app of this figure is available online at animations/6/tau. Black dots show points with the same time index that are close in the first three cases, but far apart due to overfolding in the last case

For timeseries resulting from discrete dynamical systems, $\tau = 1$ is typically a good choice, but with timeseries from continuous systems we have to pay more attention to the choice of $\tau$. Choosing too small $\tau$ leads to $R$ being stretched along the diagonal line, because the entries of the delay vector differ only slightly. This, along with noise, makes it hard to investigate the (possibly) fractal structure of the chaotic set. On the other hand, too large $\tau$ leads to overfolded reconstructions for

---

[1] Notice that now that we are back discussing practical considerations, we use $\tau$ as the delay time, which is always in multiples of the sampling time $\delta t$.

**Fig. 6.3** Autocorrelation (AC) and self-mutual information (SMI) of the $x$ timeseries of the Lorenz-63 system versus delay $\tau$. While the AC has no clear minimum, SMI does have (gray dot), yielding an optimal delay time of $\tau = 17$ (for $\delta t = 0.01$)

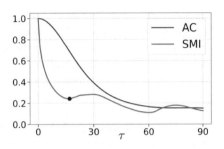

chaotic sets, a consequence of the existing stretching and folding mechanisms in the dynamics. This means that states that are far away from each other in the original set $A$ might be too close in the reconstruction $R$ or vice versa. This problem is displayed in Fig. 6.2.

The optimal value for $\tau$ is the *smallest* value where the timeseries $w$ and $w_\tau$ (with elements $w_\tau[n] = w[n + \tau]$) have minimal similarity, since you want as much new information as possible when introducing an additional dimension to the delay vector. There are different ways to estimate this optimal $\tau$. For example, you may minimize the linear correlation between $w$ and $w_\tau$ by selecting $\tau$ to be the first zero (or first minimum if no zeros) of the auto-correlation function of $w$. However, for many timeseries the auto-correlation decays exponentially with $\tau$ and therefore has no clear zero or minimum. Therefore it is in general more suitable, in our experience, to find an optimal delay time using the self-mutual information[2] of $w$ with its time-delayed image $w_\tau$. The self-mutual information is a measure of how much knowing about $w_\tau$ reduces the uncertainty about $w$ (Chap. 7) and a good choice of the delay time $\tau$ is the value where the mutual information attains its first minimum. We illustrate this in Fig. 6.3. For some timeseries, even the self-mutual information does not have a clear minimum. Then it is best to use more advanced methods, like what we describe in Sect. 6.3.3.

### 6.2.2   Choosing the Embedding Dimension

In general we want to have a value of $d$ that is as small as possible because the larger the $d$, the "thinner" the chaotic set becomes (due to finite amount of points), and this entails practical problems when one attempts to compute quantities from the reconstruction, like, e.g., the fractal dimension. A good method to estimate the smallest dimension required for a proper embedding is based on identifying so-called *approximate false nearest neighbors* (AFNN, also called *Cao's method*). This method assumes one already has an optimal $\tau$, and then computes a quantity $E_d$ for

---

[2] The mutual information is defined properly in Chap. 7 along with a sketch of how to compute it. The self-mutual information specifically has an optimized implementation in DynamicalSystems.jl as `selfmutualinfo`.

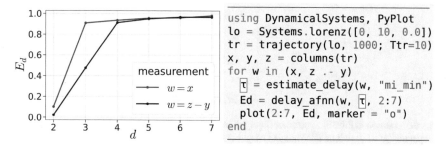

```
using DynamicalSystems, PyPlot
lo = Systems.lorenz([0, 10, 0.0])
tr = trajectory(lo, 1000; Ttr=10)
x, y, z = columns(tr)
for w in (x, z .- y)
    τ = estimate_delay(w, "mi_min")
    Ed = delay_afnn(w, τ, 2:7)
    plot(2:7, Ed, marker = "o")
end
```

**Fig. 6.4**  Using the AFNN method to find an optimal $d$ for measurements from the Lorenz-63 system. We can see that for measurements $w = x$ or $w = z - y$ the optimal $d$ is 3 and 4, respectively

an embedding with given $\tau$ and increasing $d$. The value $E_d$ starts near 0 for $d = 1$ and saturates near 1 after the optimal $d$. This is demonstrated in Fig. 6.4, also showing that different measurement functions $h$ might need higher $d$ than others.

So what is this $E_d$? First, for a given $d$ find all pairs of nearest neighbors[3] $(i, nn(i))$ $\forall i$, in the reconstructed set $R$. It is necessary here to apply the Theiler window consideration discussed in Chap. 5 when finding the neighbors. Compute the average $a^{(d)} = \langle a_i^{(d)} \rangle$ of ratios of distances of these pairs when computed in dimensions $d$ and $d + 1$

$$a_i^{(d)} = \frac{\| \mathbf{v}_i^{(d+1)} - \mathbf{v}_{nn(i)}^{(d+1)} \|}{\| \mathbf{v}_i^{(d)} - \mathbf{v}_{nn(i)}^{(d)} \|} \tag{6.3}$$

using in both dimensions the *same pairs* (uniquely characterized by their time indices). If the embedding dimension $d$ is too small, the reconstructed set is not yet properly unfolded and no diffeomorphism exists to the original set $A$. Therefore there will be some pairs of neighbors that constitute *false nearest neighbors*, as illustrated in Fig. 6.5. The value $a_i^{(d)}$ of (6.3) for these false neighbors is much larger than 1, resulting in large values of the average $a^{(d)}$. The number of false nearest neighbors is monotonically decreasing with $d$ and therefore the ratio $E_d = a^{(d+1)}/a^{(d)}$ is smaller than one, but clearly increasing as long as $d$ is too small. $E_d$ stops increasing significantly after some $d_0$, and instead converges slowly towards the value 1. This is because once a sufficiently high dimension is reached no false nearest neighbors appear anymore and $a^{(d+1)} \approx a^{(d)}$. The minimum proper embedding dimension is then $d = d_0$. Examples are shown in Fig. 6.4 for the Lorenz-63 system.

---

[3] Nearest neighbors are points in the state space that are close to each other according to some metric, typically Euclidean or Chebyshev. There are many algorithms to find such neighbors, e.g., KD-trees.

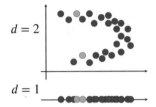

**Fig. 6.5** Demonstration of false nearest neighbors. For embedding of $d = 1$ the reconstructed set is not yet unfolded. A pair of false nearest neighbors is shown in cyan color, whose distance increases greatly in an embedding with $d = 2$

## 6.3   Advanced Delay Embedding Techniques

In this section we discuss some more advanced methods for delay embedding which are (when sensible) composable with each other.

### 6.3.1   Spike Trains and Other Event-Like Timeseries

Some timeseries represent "events" happening, for example when a neuron fires (hence the name "spike train", since the data is a sequence of spikes) or when a note is played in a music timeseries. In such timeseries one only needs to define these event times $e = \{0, 3.14, 42, \ldots\}$. One way to see this as an "actual" timeseries with respect to some time axis $t$ is to make a $w$ so that $w(t) = 1$ if $t \in e$ and $w(t) = 0$ otherwise. However using this $w$ in delay embedding is most of the time a terrible idea because of the plethora of 0 s populating the timeseries.

A better alternative is to create a new timeseries so that $s_i = e_{i+1} - e_i$, which is called the *inter-event-interval* (or *inter-spike-interval*) timeseries. For this $s$ one can make a delay embedding with delay $\tau = 1$ that for appropriate $d$ is diffeomorphic to the original chaotic set that the events were obtained from. Terms and conditions apply of course, see Further reading.

### 6.3.2   Generalized Delay Embedding

It is often the case that the physical system under study contains not one but several characteristic timescales. In such cases using the same delay time $\tau$ for all entries of the delay vector is not optimal. Instead, one should use a vector of delay times, so that each delayed entry has a different delay time, corresponding to the different timescales. In addition, if your experimental measurement provides more than one timeseries, all of them being a result of some measurement functions of the same dynamical system with the same sampling time, then you have more information at your disposal. For such a multivariate timeseries you should be able to combine some (or all) of the individual timeseries, each with its own delay, to make the most suitable embedding for your system.

We need a simple framework that can combine arbitrary timeseries with arbitrary delay times for each one. We'll call this framework *generalized delay embedding*, which is defined by a *vector* of delay times $\boldsymbol{\tau}$ and a *vector* of timeseries indices $\mathbf{j}$ (both of equal length) defining an operator $\mathcal{G}_{\mathbf{j}}^{\boldsymbol{\tau}}$ that acts on the input dataset $X$. For example, if $\mathbf{j} = (1, 3, 2)$ and $\boldsymbol{\tau} = (0, 2, -13)$ and $X = \{\mathbf{x}_1, \ldots, \mathbf{x}_N\}$ with $\mathbf{x}_i = (x_i, y_i, z_i)$, then the generalized delay vector of $X$ at index $n$ under $\mathcal{G}_{\mathbf{j}}^{\boldsymbol{\tau}}$ is

$$\mathcal{G}_{\mathbf{j}}^{\boldsymbol{\tau}} X[n] = (x[n], z[n+2], y[n-13]). \tag{6.4}$$

What is great about this framework is that it works for any imaginable scenario. For example, if you want to re-use the timeseries $x$ a second time with delay 5, you would have $\mathbf{j} = (1, 3, 2, 1)$ and $\boldsymbol{\tau} = (0, 2, -13, 5)$. There is also no ambiguity about what is the embedding dimension, because it is always the length of $\mathbf{j}$ or $\boldsymbol{\tau}$ (irrespectively of how many independent timeseries $X$ has).

### 6.3.3 Unified Optimal Embedding

So far we've only talked about how to find the optimal embedding dimension and delay time for a single input timeseries and with constant $\tau$. In addition, these methods (mutual information and AFNN) were separated from each other. As far as the embedding theorems are concerned an optimal combination $(d, \tau)$ simply provides an appropriately good coordinate system in which a diffeomorphism $F$ maps the original set $A$ to an embedded set $R$. A necessary condition is that the coordinates of the reconstruction space are sufficiently independent.

Splitting the task of finding independent coordinates into two tasks (delay time and embedding dimension) is mostly artificial and we therefore introduce in the following concepts of *unified* optimal delay coordinates based on the generalized embedding of the previous section. They try to create the most optimal (but also most minimal) generalized embedding that most accurately represents the original invariant set $A$ from which the measured timeseries $w_j$ originate from. As a result they typically yield better embeddings than the conventional approach discussed in Sect. 6.2. These methods work by coming up with a statistics that quantifies how "good" the current embedding is. Then, they proceed through an iterative process. Starting with one of the possible timeseries, additional entries are added to the generalized embedding one by one. One checks all offered timeseries $w_j$ of the measurement $X$, each having a delay time from a range of possible delay times. Then the combination of $(w_i, \tau_i)$ is picked that, when added to the (ongoing) embedding, leads to the most reduction (or increase, depending on the definition) of the statistics. This iterative process terminates when the statistics cannot be minimized (or maximized) further by adding more entries to the generalized embedding.

Okay, that all sounds great, but what's a statistic that quantifies how "good" the current embedding is? Unfortunately, describing this here would bring us beyond scope, so we point to the Further reading section. However, to give an example, one of the earliest such statistics was a quantification of how much functionally independent are the coordinates of the (ongoing) generalized embedding. Once adding an additional delayed coordinate with any delay time cannot possibly increase this independence statistic, an optimal embedding has been reached. A numerical method that performs unified delay embedding is provided in DynamicalSystems.jl as `pecuzal_embedding`.

## 6.4  Some Nonlinear Timeseries Analysis Methods

In principle, we perform nonlinear timeseries analysis throughout this book, but in the current section we list some methods that are tightly linked with delay coordinate embeddings.

### 6.4.1  Nearest Neighbor Predictions (Forecasting)

What is the simplest possible way to make a prediction (or a forecast) of a deterministic chaotic timeseries $w$ without explicitly modelling the dynamics? (Hint: it is *not* machine learning.) The simplest way will become apparent with the following thought process: Our reconstruction $R$, via delay embedding $w$, represents a chaotic set, provided we picked proper $(d, \tau)$. Assuming we satisfy the standard criteria for delay embedding (stationarity and convergence), then $R$ is invariant with respect to time: moving all initial states forward makes them stay in $R$. In addition, if we have some nearby points, when we evolve them forward in time (for an appropriately small amount of time), these points stay together in the reconstruction.

And this is where the trick lies: the last point of the timeseries $w_N$ has been embedded in $R$ as a point $\mathbf{q}$. This is our "query point". We then find the nearest neighbors of $\mathbf{q}$ in $R$ and propagate them one step forward in time. All points but the last one have an image one step forward in time already existing in $R$ so this step is trivial. Then, we make a simple claim: the average location $\tilde{\mathbf{q}}$ of the forward-images of the neighbors of $\mathbf{q}$, is the future of $\mathbf{q}$, as illustrated in Fig. 6.6. The last entry of $\tilde{\mathbf{q}}$ (or the first, depending if you did a forward or backward embedding) is the new predicted element of the timeseries $w_{N+1}$. We then embed the extended timeseries $w_{N+1}$ again in $R$ and continue using this algorithm iteratively.

## 6.4.2  Largest Lyapunov Exponent from a Sampled Trajectory

Following from the previous section, it is also easy to get a value for the largest
Lyapunov exponent given a sampled trajectory $R$ (which is typically the result of
delay embedding of a timeseries $w$). We start by finding all pairs of nearest neighbors
in $R$. Each pair $i$ has a distance $\zeta_{n=0}^{(i)}$ as in Fig. 6.6. We then go to the future image
of all points, simply by incrementing the time index by 1, and check again their
distance at the next time point: $\zeta_{n=1}^{(i)}$. Continue the process while incrementing the
time index for $k$ time steps. At each step we compute the average of the logarithms
of these distances, $\xi_n = \frac{1}{M} \sum_{i=1}^{M} \log(\zeta_n^{(i)})$ with $M = N - k$, where $N$ is the length
of the embedded timeseries.

As we have learned in Chap. 3, this $\xi_n$ can be expressed versus $n$ as $\xi_n \approx$
$n \cdot \lambda \cdot \delta t + \xi_0$ with $\delta t$ the sampling time for continuous systems and $\lambda$ the maximum Lyapunov exponent exactly as in Fig. 3.2. This happens until some maximum
threshold time $k_{\max}$ where $\xi_n$ will typically saturate due to the folding of the nonlinear
dynamics exactly like we discussed in Chap. 3. One then simply finds the slope of the
curve $\xi_n/\delta t$ versus $n$, which approximates $\lambda$. The function `lyapunov_from_data`
from DynamicalSystems.jl implements this.

## 6.4.3  Permutation Entropy

Permutation entropy connects the concepts of Chap. 5 with delay coordinates. Given
a timeseries it outputs a quantifier of complexity similar to an entropy. It works by
looking at the probabilities of relative amplitude relations of $d$ successive elements
in a timeseries. For $d = 2$ the second element can be either larger or smaller than
the previous. This gives two possibilities, and one can associate a probability to
each by counting the pairs that fall into this possibility and then dividing by the

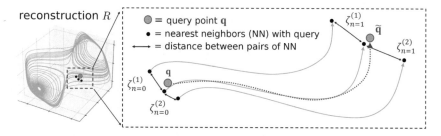

**Fig. 6.6** Demonstration of basic nonlinear timeseries analysis: nearest neighbor prediction and
Lyapunov exponent from a timeseries. Notice that the flow lines exist only to guide the eye, in the
application only the points are known

total.[4] For $d = 3$ there are 6 permutations (see Fig. 6.7), while for general $d$ the number of possible permutations equals $d!$. Each one of the possible permutations then gets associated a probability $P_i$ based on how many $d$-tuples of points fall into this permutation bin. The *permutation entropy* is simply the Shannon entropy, (5.2), of $P_i$. Sometimes $d$ is called the order of the permutation entropy.

And how does this relate to delay embedding? Well, what we do here is look at the relative amplitude relations within the points of a $d$-dimensional embedding with delay $\tau = 1$. Instead of doing a histogram of the delay embedded set with (hyper-)cubes, we partition it differently according to the possibilities of the relative amplitudes, resulting in $d!$ possible "boxes" to distribute the data in. In fact, one can extent permutation entropy to larger delays $\tau > 1$ and the approach remains valid. While permutation entropy does not have a straightforward dynamical interpretation, like, e.g., the Lyapunov exponents, it comes at the benefit of being quite robust versus noise in the data and simple to compute. Permutation entropy can also be used to detect non-stationary in the data, see the exercises.

## Further Reading

Excellent textbooks for delay coordinate embeddings and all topics related to nonlinear timeseries analysis are *Nonlinear Time Series Analysis* by Kantz and Schreiber [76] and *Analysis of Observed Chaotic Data* by Abarbanel [103], both having a practical focus and discussing real world data implications. Worth reading is also the review by Grassberger et al. [354], see also [104] for a more recent review of nonlinear timeseries analysis.

Delay coordinates embedding of timeseries was originally introduced in [105] by Packard, Crutchfield, Farmer and Shaw, stimulated by Ruelle (see [106]). It obtained an initial rigorous mathematical footing from the embedding theorem of

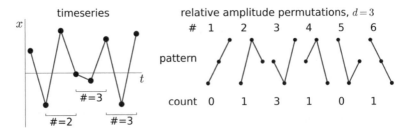

**Fig. 6.7** Permutations of relative amplitude relations for $d = 3$ based on a timeseries. Segments of length $d$ are then mapped to the corresponding permutation of relative amplitudes (also called ordinal patterns). Use the function  permentropy  from DynamicalSystems.jl to obtain the value of the permutation entropy

---

[4] In the case of equality, best practice is to choose randomly between $<$ or $>$.

Takens [107], based on Whitney [108]. That is why in the literature the delay embedding theorem is often called *Takens theorem*. However, very important generalizations of Takens' theorem came from Sauer, Yorke and Casdagli in [109, 110] and brought it in the form presented in this book, which describes the embedding condition based on the fractal dimension $\Delta$. The impact of noise on the validity of these theorems is discussed in [111]. Adaption to event-like timeseries was done by Sauer in [112]. Delay embedding theorems for forced systems were provided by Stark et al. [113, 114]. Letellier and Aguirre [115] discusses the influence of symmetries on the reconstruction problem and see also [116–118] for the impact of the observation function $h$. Because of all these contributions, we decided in this book to drop the term "Takens theorem" and use the term "delay embedding theorem" instead.

For signals $w(t)$ from continuous systems, as an alternative to delay coordinates, *derivative coordinates* $\mathbf{v}(t) = (w(t), \dot{w}(t), \ddot{w}(t), \ldots)$ can be used for state space reconstruction. This approach is justified by embedding theorems similar to those for delay coordinates [107, 109]. Since estimating higher temporal derivatives from noisy (measurement) data can be very challenging, this method of state space reconstruction is suitable for low dimensional reconstructions, only. An alternative method to overcome or bypass the issue of selecting a proper time delay was introduced by Broomhead and King [119], combining embedding in a very high dimensional space in combination with linear dimensional reduction (principal component analysis).

The references regarding the conventional approach to optimal embedding are too numerous to list all of them here. Important to point out is the work of Kennel, Brown and Abarbanel [120], which was the first to come up with a way to estimate an optimal embedding dimension based on the concept of false nearest neighbors. Then Cao [121] provided an improved method to do the same. More historical references and reviews of the conventional approach to delay embeddings and can be found in [111, 122]. Details on embeddings of event-like timeseries can be found in [112]. The impact of linear filtering a signal prior to delay embedding has be investigated by Broomhead et al. [123] and Theiler and Eubank [124].

Unified approaches to delay coordinate reconstructions have been developed only recently, starting with the work of Pecora et al. [122] that we described loosely in Sect. 6.3.3 (which, by the way, also contains excellent historical overview of delay embeddings). Further improvements have been proposed since then, see, e.g., Krämer et al. [125] and references therein. All unified approach based methods (to date) are implemented in DynamicalSystems.jl.

The framework of nearest neighbor prediction described in Sect. 6.4.1 is implemented in the software TimeseriesPrediction.jl. This framework can also be used to predict the dynamics of spatially extended systems (Chap. 11), see, e.g., [126]. Permutation entropy, originally introduced by Bandt and Pompe [127], has been extended in various ways [128–130].

## Selected Exercises

6.1 Write your own code that can delay embed timeseries. Ensure that it is general enough to allow both embedding with a constant delay time $\tau$ but also the generalized embedding of Sect. 6.3.2. Then produce $x, y, z$ timeseries of the Lorenz-63 system (1.5) with sampling time $\delta t = 0.01$. Let $\tau_x = 17$ and $\tau_z = 15$. Create three dimensional generalized embeddings and plot them, using the combinations: (i) $\mathbf{j} = (1, 1, 1)$ and $\boldsymbol{\tau} = (0, \tau_x, 2\tau_x)$, (ii) $\mathbf{j} = (3, 3, 3)$ and $\boldsymbol{\tau} = (0, \tau_z, 2\tau_z)$, (iii) $\mathbf{j} = (1, 1, 3)$ and $\boldsymbol{\tau} = (0, \tau_x, \tau_z)$, (iv) $\mathbf{j} = (2, 3, 3)$ and $\boldsymbol{\tau} = (0, \tau_z, 2\tau_z)$. Which case does not resolve the two "wings" of the Lorenz-63 attractor, and why?

6.2 Load the timeseries from datasets 1, 2, 6 and embed them in $d = 3$-dimensional space with constant delay time $\tau$. Use the mutual information as a way to obtain optimal delay time $\tau$. How do the reconstructed sets change when changing $\tau$?

6.3 Implement the method of false nearest neighbors described in Sect. 6.2.2. To find nearest neighbors in higher dimensional spaces, you can still write a simple brute-force-search that compares all possible distances. However it is typically more efficient to rely on an optimized structure like a KDTree. Simulate a timeseries of the towel map (e.g., pick the first coordinate)

$$
\begin{aligned}
x_{n+1} &= 3.8x_n(1 - x_n) - 0.05(y_n + 0.35)(1 - 2z_n) \\
y_{n+1} &= 0.1\left((y_n + 0.35)(1 - 2z_n) - 1\right)(1 - 1.9x_n) \\
z_{n+1} &= 3.78z_n(1 - z_n) + 0.2y_n
\end{aligned}
\tag{6.5}
$$

and from it try to find the optimal embedding dimension. Compare your code with the function `delay_afnn` from DynamicalSystems.jl to ensure its accuracy. *Hint: since this is a discrete system, $\tau = 1$ is the delay time of choice.*

6.4 Simulate a timeseries of the Rössler system, (4.1), for some sampling time $\delta t$. Using the auto-correlation function or self-mutual information (we strongly recommend to use the mutual information version), estimate the optimal embedding time $\tau$ for your timeseries. How do you think this should depend on $\delta t$? Re-do your computation for various $\delta t$ and provide a plot of $\tau(\delta t)$.

6.5 Load timeseries from exercise datasets 1, 3, 5, 12 (and for multidimensional data use the first column). For each try to achieve an optimal delay coordinates embedding using the conventional approach.

6.6 Repeat the previous exercise for datasets 6, 9, 10, 13.

6.7 Implement the method of Sect. 6.4.2 for computing the maximum Lyapunov exponent from a given dataset. Apply the result to dataset 2, which has a sampling time of 0.05. Compare your code with the function `lyapunov_from_data` from DynamicalSystems.jl to ensure its accuracy.

6.8 Repeat the previous exercise for datasets 3, 6, 10. The sampling times are 0.01, 1, 0.01, respectively.

6.9 How do you expect the fractal dimension of the embedded sets to behave as you increase the embedding dimension $d$? Test your hypothesis using datasets 3, 6 and 12.

6.10 Simulate a $N = 10,000$-length timeseries of the Hénon map, (3.3), and record the first variable as a timeseries. Attempt a prediction of $n = 100$ time steps using the algorithm of the simple nearest-neighbour-based forecasting of Sect. 6.4.1.

6.11 Quantify the dependence of the error of your prediction on $N$ by calculating the normalized root mean squared error given by:

$$\text{NRMSE}(x, y) = \sqrt{\frac{\text{MSE}(x, y)}{\text{MSE}(x, \bar{x})}}, \quad \text{MSE}(x, y) = \frac{1}{n} \sum_{i=1}^{n} (x_i - y_i)^2$$

where $x$ is the true continuation of the timeseries to-be-predicted, $y$ is the prediction and $\bar{x}$ the mean of $x$. For a given $N$, record the value $n_c$ where the NRMSE exceeds 0.5 ($n_c$ is related to the prediction horizon, Sect. 3.1.2). How would you expect that $n_c$ depends on $N$? Make a plot and confirm your hypothesis.

6.12 Repeat the above exercise for both the $x$ and $z$ timeseries from the chaotic Rössler system, (4.1) with sampling $\delta t = 0.1$. Which one would you expect to be predicted easier (in the sense of having larger $n_c$ for a given $N$)? *Hint: take into account the Theiler window when finding nearest neighbors!*

6.13 Load dataset 7, which is a non-stationary timeseries with observational noise. Use the permutation entropy (write your own function to compute it!) and report its value for various orders $d$. Compare the results with the function permentropy from DynamicalSystems.jl.

6.14 Analyze the timeseries of the previous exercise by splitting it into windows of length $W$ each with overlap $w < W$. For each window calculate permutation entropies of different orders, and plot all these results versus time. Can you identify time-windows of characteristically different behavior from these plots? How do your results depend on $W$, $w$ and the order of the permutation entropy? In the literature, this process is called *identifying regime shifts*, and the permutation entropy is one of several tools employed to find them.

# Chapter 7
# Information Across Timeseries

**Abstract** In Chap. 5 we have discussed how to calculate several variants of information contained within a timeseries or set. But an equally interesting question is
how much information is shared between two different timeseries. Answering this
allows us to say how much they are "correlated" with each other, which is useful
in practical applications. It is also possible to take this a step further and ask about
information transfer (i.e., directed information flow) between timeseries, which can
become a basis for inferring connections between sub-components of a system. This
is used in many different fields, as for example neuroscience. We close the chapter by
discussing the method of surrogates, which is used routinely in nonlinear dynamics,
e.g., for identifying nonlinearities in measured timeseries.

## 7.1 Mutual Information

There are several ways to measure correlation between two timeseries $x$ and $y$.
Typically used examples are the *Pearson correlation coefficient*, which measures
linear correlation between $x$ and $y$, and the *Spearman correlation coefficient*, which
can measure arbitrary nonlinear but only monotonic relations. Both coefficients are
bounded in $[-1, 1]$, with $\pm 1$ meaning perfect (anti-)correlation and 0 no correlation. However these coefficients are not well suited for non-monotonic relations or
delayed interactions. For example, if $x(t)$ is a timeseries fluctuating between $-1$
and 1 and $y(t) = x(t)^2 + \varepsilon(t)$ with noise $\varepsilon$. Another example is comparing the $y(t)$
and $z(t)$ timeseries from the chaotic Lorenz system (1.5) introduced in Sect. 1.1.1.
For both of these examples, the aforementioned correlation coefficients would be
either approximately zero even though both examples are in some sense "perfectly
correlated".

An established method that estimates general correlation between $x$ and $y$ is their
mutual information (MI)

G. Datseris and U. Parlitz, *Nonlinear Dynamics*, Undergraduate Lecture Notes in Physics,
https://doi.org/10.1007/978-3-030-91032-7_7

$$MI(x, y) = \sum_n \sum_m P_{xy}(x_n, y_m) \log \left( \frac{P_{xy}(x_n, y_m)}{P_x(x_n) P_y(y_m)} \right)$$
$$= H(x) + H(y) - H(x, y) \geq 0 \qquad (7.1)$$

where $H(x)$ and $H(y)$ are the Shannon entropies[1] (5.2) of $x$ and $y$, respectively, based on the amplitude binning definition of $P$ as in Sect. 5.2.1. $H(x, y)$ is their joint entropy, which is simply the entropy of the set $A = \{(x_i, y_i) : i = 1, \ldots, N\}$.

Intuitively MI measures how much knowing one of these observables reduces uncertainty about the other. The higher the MI, the more information $x$ and $y$ share. This can also be seen mathematically, since the formula $\sum_x P(\mathbf{x}) \log (P(\mathbf{x})/Q(\mathbf{x}))$ (called the Kullback–Leibler divergence) measures how much different a probability distribution $P$ is from a reference distribution $Q$. In our case we measure the difference of the joint probability of $P_{xy}$ over the independence case $Q = P_x P_y$ (thus $MI = 0$ means that $P_{xy} = Q = P_x P_y$, i.e., independence). The distributions $P_{xy}, P_x, P_y$ can be calculated for example with the amplitude binning method (histogram) we discussed in Sect. 5.2 and then be inserted into the Shannon formula.

The actual value of the MI, $m = MI(x, y)$, is a measure of information, like the Shannon entropy of Sect. 5.2. How can one use $m$ to evaluate whether $x$, $y$ are correlated? A simple way to do this is to check how $m$ compares with the so-called *null hypothesis*, "$x$ and $y$ are uncorrelated". To estimate that, we first randomly shuffle (i.e., randomly permute) the sequences $x$ and $y$ and compute (7.1) again. The shuffled versions of $x$, $y$ are called *shuffled surrogates* (see Sect. 7.4 for more on surrogate data). But why shuffle $x$, $y$? Well, these shuffled surrogates do not alter the distributions $P_x$, $P_y$ but *may* alter $P_{xy}$. If the data were originally uncorrelated and $P_{xy} = P_x P_y$, then the shuffled surrogates will have the same MI as the original data.

To test this, we create shuffled surrogates for let's say 10,000 times, and the resulting 10,000 MI values we calculate give a distribution (asymptotically Gaussian for large $N$). This process is called *bootstrapping*. The further away $m$ is from the mean of that distribution, $\mu_{null}$, the more unlikely the null hypothesis is. Typically one calculates the standard deviation of this distribution $\sigma_{null}$ and claims significant correlations if $m$ deviates from the mean by more than $2\sigma_{null}$, since the interval $\mu_{null} \pm 2\sigma_{null}$ is the 95% confidence interval assuming a Gaussian distribution (using $3\sigma$ gives 99.73% confidence).

We demonstrate this process in Fig. 7.1 by using two timeseries of the logistic map, $x$, $y$ where $y_n = x_{n+1}$. We know that by definition $x$, $y$ are connected by the rule of the logistic map, i.e. $y_n = rx_n(1 - x_n)$. We further add noise to $x$, $y$ to make this application a bit more realistic. Notice that for these timeseries, both Pearson and Spearman coefficients are $\approx 0$ but the MI succeeds in showing that these two timeseries are very much correlated.

---

[1] Notice that other methods of defining the probabilities $P$, such as those discussed in Chap. 5 or Sect. 6.4.3 can be used to define MI, see exercises.

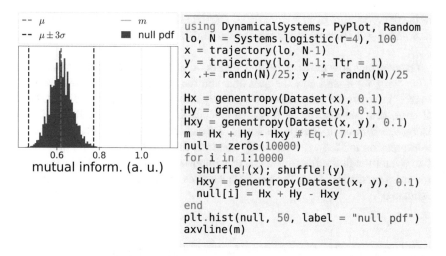

```
using DynamicalSystems, PyPlot, Random
lo, N = Systems.logistic(r=4), 100
x = trajectory(lo, N-1)
y = trajectory(lo, N-1; Ttr = 1)
x .+= randn(N)/25; y .+= randn(N)/25

Hx = genentropy(Dataset(x), 0.1)
Hy = genentropy(Dataset(y), 0.1)
Hxy = genentropy(Dataset(x, y), 0.1)
m = Hx + Hy - Hxy # Eq. (7.1)
null = zeros(10000)
for i in 1:10000
    shuffle!(x); shuffle!(y)
    Hxy = genentropy(Dataset(x, y), 0.1)
    null[i] = Hx + Hy - Hxy
end
plt.hist(null, 50, label = "null pdf")
axvline(m)
```

**Fig. 7.1** Demonstration of estimating correlation via the mutual information between two noisy timeseries $x$ and $y$ obtained from the logistic map. The code calculates entropies by binning in boxes of size 0.1, and takes advantage of the fact that $H(x)$, $H(y)$ do not change value if we shuffle $x$, $y$ around, only $H(x, y)$ does

## 7.2  Transfer Entropy

Mutual information treats the timeseries statistically and it is also invariant under swapping $x$ with $y$. Thus, it cannot contain directional information or distinguish past from future. One is often interested in flow of information *from* one timeseries *to* another. This is a directional relation, and since flow of information occurs with finite speed, such a relation must necessarily relate the past and present of one timeseries with the future of another. It is possible to modify the MI by adding a time delay in one of two timeseries (which we have already taken advantage of in Chap. 6), however there are better alternatives for quantifying *transfer* of information.

One is the *transfer entropy* (TE), which is used more and more in many diverse areas, as in, e.g., neuroscience to identify which neurons are interacting with which or to estimate social influence in social networks. It is given by

$$TE_{y \to x} = \sum_{\mathbf{x}_n, \mathbf{y}_n, x_{n+\ell}} P(x_{n+\ell}, \mathbf{x}_n, \mathbf{y}_n) \log \left( \frac{P(x_{n+\ell} | \mathbf{x}_n, \mathbf{y}_n)}{P(x_{n+\ell} | \mathbf{x}_n)} \right) \tag{7.2}$$

where

$$\mathbf{x}_n = \mathcal{D}_{d_x}^{-\tau_x} x[n] = (x[n], x[n - \tau_x], \ldots, x[n - \tau_x(d_x - 1)])$$
$$\mathbf{y}_n = \mathcal{D}_{d_y}^{-\tau_y} y[n] = (y[n], y[n - \tau_y], \ldots, y[n - \tau_y(d_y - 1)])$$

are the delay vectors of $x$, $y$ from (6.1), $P(A|B)$ is the conditional probability of $A$ on $B$ and $\ell \geq 1$ (summation is triple: over embedded states $\mathbf{x}_n$, $\mathbf{y}_n$ and future state $x_{n+\ell}$).

TE quantifies how much *extra* information we gain for the future of one timeseries $x$ (the "target") if we know the past of $x$ and the past of another timeseries $y$ (the "source"), *versus* the information we gain by only knowing the past of $x$. Like the MI case, this can also be seen mathematically in terms of the Kullback–Leibler divergence for conditional probabilities. $TE_{y \to x}$ is measured in bits. As expected, the expression for $TE_{y \to x}$ necessarily connects the "past" of $x$, $y$ with the "future" of $x$. Even though the future is used, it is important to understand that TE does no timeseries prediction. Only the available samples of $x$, $y$ are used, no new ones are "estimated".

### 7.2.1  Practically Computing the Transfer Entropy

Here we will use the software TransferEntropy.jl, that has several different estimators for the TE, so we can go directly into discussing the practical difficulties one is faced with when attempting to estimate TE. The first difficulty is choosing appropriate embedding parameters. Although techniques discussed in Chap. 6 can help, for TE related applications the so-called Ragwitz criterion is often a better choice. According to this, one aims to find the pair $(d, \tau)$ that minimizes the error of the nonlinear timeseries prediction technique of Sect. 6.4.1 (for each $x$, $y$ individually). Then one has to choose $\ell$. Values greater than 1 (only $+1$ is included in the original definitions of TE) target real world scenarios where the timescale of information transfer is not known a priory. In these cases one can compute TE for several $\ell$ and pick the maximum TE value, or even average the results.

Besides choosing the delay parameters, one also has to choose the process by which to estimate the probabilities $p$ in (7.2). Basic estimators are in principle based on the amplitude binning or nearest neighbors mentioned in Chap. 6, both of which require a choice the size scale $r$. The outcome of the computation depends on $r$, and of course also depends on the chosen estimator. To demonstrate how important this is, in Code 7.1 we calculate the TE between components of 100 coupled Ulam maps

$$x_{n+1}^{(m)} = f(\epsilon x_n^{(m-1)} + (1-\epsilon)x_n^{(m)}); \quad f(x) = 2 - x^2 \tag{7.3}$$

with $\epsilon$ the coupling strength, $m$ runs from 1 to 100, and the maps are coupled circularly, i.e., $m = 0$ is considered the same as $m = 100$. We show the results of transfer entropy between $x^{(1)}$ and $x^{(2)}$ in Fig. 7.2. Theoretically, due to the local coupling in (7.3), there should be information flow from $x^{(m-1)}$ to $x^{(m)}$ and the opposite direction should have (almost[2]) zero TE, because the equation for $x^{(m)}$ involves $x^{(m-1)}$ but

---

[2] Besides the unidirectional local coupling there is also the global coupling "around the circle" that results in a very weak impact of $x^{(m)}$ on $x^{(m-1)}$.

the one for $x^{(m-1)}$ does not involve $x^{(m)}$. For $\epsilon = 0$ both directions should have zero TE and for $\epsilon \approx 0.18$ and $0.82$ again both directions should have zero TE because the system has a periodic window there (and thus the past of the target variable is enough to determine its future).

As we know that this is a discrete map, we can choose $\tau = 1, d = 2$ for both $x$, $y$, and $\ell = 1$, leaving only $r$ to "tune". As is evident in Fig. 7.2, for inappropriate (i.e., too small) choices of the bin size $r$ like $0.01$ one can get almost equivalent results for both directions and also see artifacts like the TE not being zero for zero coupling. What happens in this case (and this of course depends on the length of the timeseries) is that the boxes of the histograms used to estimate the probabilities in (7.2) are all either empty or have only one (or at most a few) points, giving almost identical probability values for any box. For better choices of $r$ the theoretically correct direction has much higher TE. In general, better but more involved ways exist for computing TE, based on probability estimators not using amplitude binning, see Further reading.

A final "gotcha" about transfer entropy, is that it requires a lot of data points. Even in the trivial example above with the coupled Ulam maps, we are creating a 5-dimensional embedded space (1 future value of $x$, two past values of $x$ and two past values of $y$). The higher the dimension of the embedded space, the more data points one needs to fill it sufficiently densely in order to get accurate estimates of all the probabilities $P$ that go into (7.2).

**Code 7.1** Calculation of the transfer entropy between first two components of coupled Ulam maps 7.3 using TransferEntropy.jl.

```julia
using DynamicalSystems, TransferEntropy
ds = Systems.ulam(100)
genmeth(r) = VisitationFrequency(RectangularBinning(r))
methods = genmeth.((0.01, 0.1, 0.4))
rs = 0.0:0.01:1.0
tes = [zeros(length(rs), 2) for j in 1:length(methods)]

for (i,r) in enumerate(rs), (j,meth) in enumerate(methods)
    set_parameter!(ds, 1, r)
    tr = trajectory(ds, 10000; Ttr = 10000)
    X1 = tr[:, 1]; X2 = tr[:, 2]
    # default values for τ, d are 1 in `transferentropy`
    tes[j][i, 1] = transferentropy(X1, X2, meth)
    tes[j][i, 2] = transferentropy(X2, X1, meth)
end
```

## 7.2.2  Excluding Common Driver

A case to worry about is if both $x$ and $y$ have a common driver, let's say $z$. In such scenarios $TE_{y \to x}$ can be significantly non-zero, even though we know that there is

**Fig. 7.2**  Transfer entropy for coupled Ulam maps (see Code 7.1), for various bin sizes $r$. Solid lines are TE from $x^{(1)}$ to $x^{(2)}$, dashed lines from $x^{(2)}$ to $x^{(1)}$

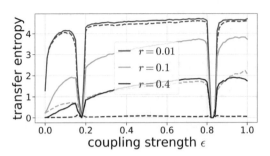

no real information transfer from $y$ to $x$, but the non-zero value is instead an artifact of $z$ driving both $x$, $y$. To account for this, one can condition the probabilities of (7.2) to the third timeseries $z$, making a *conditional* TE

$$TE_{y \to x|z} = \sum_{\substack{x_{n+\ell}, \mathbf{x}_n, \\ \mathbf{y}_n, \mathbf{z}_n}} P\left(x_{n+\ell}, \mathcal{D}_{d_x}^{-\tau_x} x[n], \mathcal{D}_{d_y}^{-\tau_y} y[n], \mathcal{D}_{d_z}^{-\tau_z} z[n]\right) \times$$

$$\log\left(\frac{P\left(x_{n+\ell} | \mathcal{D}_{d_x}^{-\tau_x} x[n], \mathcal{D}_{d_y}^{-\tau_y} y[n], \mathcal{D}_{d_z}^{-\tau_z} z[n]\right)}{P\left(x_{n+\ell} | \mathcal{D}_{d_x}^{-1} x[n], \mathcal{D}_{d_z}^{-\tau_z} z[n]\right)}\right). \qquad (7.4)$$

If $z$ already gives all extra information for the future of $x$, then $y$ provides nothing further and $p(x|x, y, z) \approx p(x|x, z)$ in the above equation. Then, the conditional TE between $x$ and $y$ will be approximately zero in this case.

## 7.3  Dynamic Influence and Causality

At first it seems that TE encapsulates a "causality" definition, but one should be careful with such statements. In fact, even the conceptual definition of causality can vary wildly depending on the application scenario and has remained strongly controversial in the scientific community until now. TE hints at *effective* connections (also known as *functional connectivity*), which may or may not be causal. Besides, in the real world, it is almost never the case that TE in one direction is nonzero and in the other zero. It could be that the direction of higher TE indicates the absolute causal flow, but it is much more likely that it indicates the dominant information flow direction, as real systems are complex with multiple interacting components. In addition, the two timeseries participating in TE may represent dynamics at different timescales or dynamic variables with different information content, both of which can cause asymmetries in the TE.

Nevertheless, TE is one of the many methods employed in *causal inference*, a sub-field of nonlinear timeseries analysis that aims to uncover connections between timeseries. Sometimes it is best to not think in terms of causality, but instead *dynamic influence*, in the sense of one component of a system being, e.g., a driver of some other component, or participating in the other component's equation(s). It is in fact exceptionally useful to analyze timeseries in search of such dynamic influence. This analysis can be helpful in understanding the connections between the components of the system and to find out which ones are the most important. This can guide the modelling of a real system. For example, we mentioned that TE measures effective connections, and sometimes information about such functional connectivity is by itself useful enough (or even more useful) than knowing the exact physical connections. When one is treating a high dimensional system with many connections (e.g., a neural system), it is useful to reduce it to less components and less connections, by typically keeping the components that are most connected, in the dynamic sense. TE provides this knowledge and allows one to do a low-dimensional effective representation of the high dimensional system.

## 7.3.1 Convergent Cross Mapping

A different method that has been used to infer dynamic connections, based heavily on the ideas of Chap. 6, is the *convergent cross mapping* (CCM). The method takes a dynamical systems approach, and claims that two timeseries $x$, $y$ are dynamically connected if they are both variables of the same underlying dynamical system. Because of the delay embedding theorems discussed in Chap. 6, we can numerically uncover this connection. Delay embedding $x$ and $y$ (with proper parameters) makes two sets $M_x$, $M_y$, which are diffeomorphic to the original attractor (or in general invariant set) of the real system. Therefore there must be a connection between $M_x$, $M_y$ as well, and the task is to uncover it.

The actual implementation of CCM is rather simple (see also Code 7.2), and measures dynamic influence of $x$ on $y$ (which in general is not the same for $y$ to $x$, as one component may have more influence over the dynamics of the other than the inverse scenario). Assume both timeseries are synchronously recorded and have length $N$. First, $x$ is delay embedded, as instructed by Chap. 6, leading to a $d$-dimensional set $M_x$. Then, similarly with the nonlinear timeseries prediction in Sect. 6.4.1, we find the $d + 1$ nearest neighbors of the $i$th point $\mathbf{x}_i$ of $M_x$ (this is done for all $i$). We want to see if the indices of these nearest neighbors map to indices of nearest neighbors around the $i$th point of the embedding of $y$, $\mathbf{y}_i \in M_y$, as illustrated in Fig. 7.3. If they do, then it is likely that there is a connection from $x$ to $y$ and thus the two timeseries might be coming from the same dynamical system. To check for a connection from $y$ to $x$ you repeat the process with the sets exchanged.

We can compute CCM from $x$ to $y$ without even embedding $y$, by making a new timeseries $\psi$ based on the indices of nearest neighbors found in $M_x$. For each $i$ let $nn(i)$ be the indices of the $d + 1$ nearest neighbors of $\mathbf{x}_i$. Then define $\psi_i$ as follows

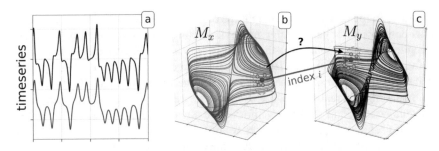

**Fig. 7.3** Demonstration of the idea of convergent cross mapping (CCM). Two timeseries $(x, y$ components of the Lorenz-63 system) are reconstructed in 3D space via delay coordinates. The task is to see if nearest neighbors of point $\mathbf{x}_i \in M_x$ map to neighbors around point $\mathbf{y}_i \in M_y$

$$\psi_i = \sum_{j \in nn(i)} w_j y_j, \quad w_j = u_j / \sum_{j \in nn(i)} u_j, \quad u_j = \exp\left(-\frac{||\mathbf{x}_i - \mathbf{x}_j||}{\max_j\{||\mathbf{x}_i - \mathbf{x}_j||\}}\right).$$

(7.5)

The measure of CCM is any measure that quantifies how much $\psi$ deviates from $y$ and typically the Pearson correlation coefficient $r$ is used. $r$ reaches the value 1 for perfect dynamic influence of $x$ on $y$.

**Code 7.2** Convergent Cross Mapping implementation.

```
using DynamicalSystems, Neighborhood
using LinearAlgebra, Statistics
function ccm(x, y, d, τ)
  ψ = copy(y); Mx = embed(x, d, τ); tw = Theiler(τ);
  # bulk search of d+1 neighbors for all i points:
  idxs=bulkisearch(KDTree(Mx), Mx, NeighborNumber(d+1), tw)
  for i in 1:length(Mx)
    nni = idxs[i]; xi = Mx[i]; n1 = norm(xi - Mx[nni[1]])
    # implement equation (7.5):
    u = [exp(-norm(xi - Mx[j])/n1) for j ∈ nni]
    w = u ./ sum(u)
    ψ[i] = sum(w[k]*y[j] for (k, j) in enumerate(nni))
  end
  return cor(y, ψ) # Pearson correlation coef.
end
```

That's great, but we still need a way to see if the number outputted by the CCM framework is significant, and how to distinguish between who influences whom more, if possible. Provided that indeed $x, y$ come from the same underlying deterministic dynamics, a nice feature of CCM is that $r$ *converges* to a saturation value as the timeseries length increases. Convergence happens because the reconstructed sets $M_x, M_y$ are filled more densely and (7.5) becomes more accurate. Thus, a simple test is to calculate $r$ for increasing subsets $N_s$ of the given timeseries from, e.g.,

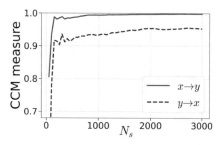

**Fig. 7.4** Result of applying the CCM method for $x$ and $y$ timeseries from the Lorenz-63 system, using Code 7.2, for various subsets of size $N_s$ of the total timeseries

$N_s = N/100$ to $N$ and plot $r(N_s)$ and see if it converges (the curve saturates to a plateau). This is shown in Fig. 7.4. In the limit of infinite amount of available data $N \to \infty$, CCM must converge to 1 if there is dynamic influence, but of course in reality finite $N$ will (typically) make it converge to a value smaller than 1. If $r_{x \to y}$ converges faster than $r_{y \to x}$, then $x$ may have more influence on $y$. In this case, and for finite data, $r_{y \to x}$ will likely also converge to a smaller value than $r_{x \to y}$ (if it manages to converge). Be aware that there can be situations where two $x, y$ timeseries may come from the same system, but CCM fails to identify this, because $x$ or $y$ may be incapable of yielding proper reconstructions of the dynamics, as we discussed in Sect. 6.1.2. For timeseries from strongly coupled systems synchronization may prevent detection of coupling directions as illustrated in Sect. 9.3.2.

## 7.4 Surrogate Timeseries

The method of surrogates is a powerful technique that is used primarily for two reasons: (1) to test whether a given timeseries does, or doesn't, satisfy a specific "hypothesis" and (2) to test whether an interaction exists between two or more given timeseries. Case (2) is connected with the ideas already discussed in the previous sections. Here we will discuss application (1), because it is the main way one can distinguish deterministic dynamics from noise. This is an important tool for evaluating, e.g., estimates of fractal dimensions.

For example, the first surrogate methods were created to test the hypothesis, whether a given timeseries $x$ that appears noisy is actually coming from a linear stochastic process or not. If not, it might be from a nonlinear dynamical system with measurement noise. But how does one "test" whether the underlying hypothesis is satisfied or not? The process is the same with giving a confidence interval for the mutual information, Sect. 7.1. First, you have to specify a null hypothesis you want to test against, for example "the timeseries follows a linear stochastic process". Based on the hypothesis you start generating *surrogate timeseries*: fake timeseries created by manipulating the original timeseries in a way that conserves the defining properties of the chosen null hypothesis. In our example, we would need to generate timeseries that have identical power spectrum as the original timeseries, because the power spectrum is the defining property of a linear stochastic process. Thus, each surrogate method is suited to test a specific null hypothesis.

Then, you choose a discriminatory statistic $q(x)$ and apply it to the input timeseries $x$ and to several different surrogate realizations $s$. The latter give a distribution of $q$ values for timeseries that obey the hypothesis. Well, if the discriminatory statistic $q(x)$ of the input value is significantly "outside" the obtained distribution (within some confidence interval), one rejects the hypothesis. If not, the hypothesis cannot be rejected. Such a "rejection" of this hypothesis provides a hint that the origin of $x$ may be a nonlinear dynamical system, possibly with added uncorrelated noise. Thus, surrogate methods do not tell you what $x$ can be, they only tell you what it *can't be*, since other reasons why the timeseries does not satisfy the null hypothesis may exist, outside deterministic nonlinearity.

The discriminating statistic $q$ depends on the chosen null hypothesis. Some options are the correlation dimension, the auto-mutual-information (the mutual information of a timeseries with itself shifted forwards for one or more steps), or the time-series time-irreversibility measure given by $a = \frac{1}{N} \sum (x_{n+1} - x_n)^3$. The value of $a$ is near zero for time-reversible timeseries, which holds for timeseries from all linear stochastic systems and conservative nonlinear systems. Thus, $a$ can be used as a test when searching for dissipative nonlinearity. Furthermore, it is always important to choose a statistic that is sufficiently robust versus noise and can be calculated sufficiently accurately given the amount of data in $x$ (for example, Lyapunov exponents estimated from timeseries are typically not robust versus noise).

### 7.4.1   A Surrogate Example

But why is the surrogate method necessary at all? Shouldn't it be "obvious" whether a timeseries is noise or not just by looking at it? Nope! We have already seen in the discussion around Fig. 1.1 that timeseries of discrete systems can appear random. This is also true for sparsely sampled continuous systems, and to make things worse in reality there is always a small amount of noise in measured timeseries. Besides, linear stochastic processes can support oscillations and other kinds of behavior. Such processes are defined on the basis of autoregressive moving-average processes, called ARMA model, with defining equation

$$x_n = \xi_n + \sum_{i=1}^{p} \phi_i x_{n-i} + \sum_{i=1}^{q} \theta_i \xi_{n-i} \tag{7.6}$$

where $\phi_i, \theta_i$ are fixed parameters and the elements $\xi_i$ are independent, normally distributed random numbers. ARMA models are linear because $x_n$ depends linearly on previous terms, and they are stochastic, instead of deterministic, because of the inclusion of $\xi$.

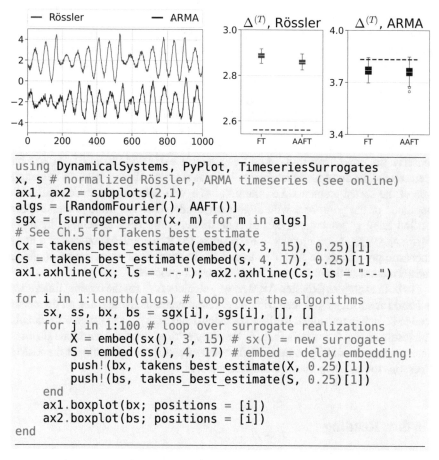

```
using DynamicalSystems, PyPlot, TimeseriesSurrogates
x, s # normalized Rössler, ARMA timeseries (see online)
ax1, ax2 = subplots(2,1)
algs = [RandomFourier(), AAFT()]
sgx = [surrogenerator(x, m) for m in algs]
# See Ch.5 for Takens best estimate
Cx = takens_best_estimate(embed(x, 3, 15), 0.25)[1]
Cs = takens_best_estimate(embed(s, 4, 17), 0.25)[1]
ax1.axhline(Cx; ls = "--"); ax2.axhline(Cs; ls = "--")

for i in 1:length(algs) # loop over the algorithms
    sx, ss, bx, bs = sgx[i], sgs[i], [], []
    for j in 1:100 # loop over surrogate realizations
        X = embed(sx(), 3, 15) # sx() = new surrogate
        S = embed(ss(), 4, 17) # embed = delay embedding!
        push!(bx, takens_best_estimate(X, 0.25)[1])
        push!(bs, takens_best_estimate(S, 0.25)[1])
    end
    ax1.boxplot(bx; positions = [i])
    ax2.boxplot(bs; positions = [i])
end
```

**Fig. 7.5** Surrogate test with Takens' best estimator $\Delta^{(T)}$ as the discriminatory statistic, for a Rössler timeseries (4.1 with parameters $a = 0.165$, $b = 0.2$, $c = 10.0$) and an ARMA timeseries as in the text. Timeseries are shown on the left, the right plots display results of Takens' best estimator (dashed lines are values obtained for the original timeseries, while the boxes are surrogate values, where the whiskers extent to the 5–95% quantiles of the distribution). Embedding dimension and delay ((3, 15) for Rössler, (4, 17) for ARMA) were chosen according to the methods of Chap. 6. Equation (5.8) is used to calculate $\Delta^{(T)}$

To demonstrate how hard it is to visually distinguish such timeseries from ones coming from a nonlinear chaotic system, let's simulate the Rössler system from (4.1). In Fig. 7.5 we show the evolution of the $x$ variable which undergoes chaotic oscillations. The $x$-signal was sampled with a sampling time of 0.1 and then measurement noise with an amplitude of 10% was added. We compare this with an ARMA model with $\phi = (1.625, -0.284, -0.355)$ and $\theta = (-0.96)$. The similarity is striking!

To identify nonlinearity we formulate a null hypothesis test using the Takens' best estimator $\Delta^{(T)}$ (Sect. 5.4) as the discriminatory statistic $q$. For generating surrogates, we will use surrogate methods that satisfy the null hypothesis of a linear stochastic process. Such processes are completely described by their autocorrelation function, or equivalently, by their power spectrum (Wiener–Khinchin theorem). Thus we choose methods whose generated surrogates $s$ have the same power spectrum as $x$. Specifically, we will use two common methods: the random phase Fourier transform (FT) and the amplitude adjusted Fourier transform (AAFT).

FT is quite simple: it performs the Fourier transform of the input $x$, randomly shuffles the phases of the Fourier components, and then transforms it back to generate the surrogate $s$. This does not conserve the amplitude distribution of $x$ but does conserve the power spectrum. AAFT is a simple extension to the FT. After $s$ has been created, its amplitude distribution is equated to that of $x$ by equating the amplitude of the sorted elements. I.e., the point of $s$ with largest amplitude obtains the amplitude of the point of $x$ with largest amplitude. Then the same process repeats for 2nd largest, 3rd largest, etc. In code this is really easy to achieve by doing `s[sortperm(s)] .= sort(x)`. FT is suited to test the null hypothesis of a linear stochastic process while AAFT addresses the hypothesis of a monotonic nonlinear transformation of the output of a linear stochastic process.

For both examples (Rössler, ARMA) we calculate $\Delta^{(T)}$ for the original timeseries, and also for surrogates created with the different methods. Notice that all timeseries need to be delay-embedded to allow for calculation of $\Delta^{(T)}$, see Chap. 6 for details. The result of Fig. 7.5 shows that for the Rössler timeseries the surrogate data test clearly *rejects* the null hypothesis of a linear stochastic process, while for the ARMA timeseries the null hypothesis *cannot* be rejected.

## Further Reading

More details about estimating correlations using mutual information are given in [131]. The transfer entropy was introduced by Schreiber [132] and Paluš et al. [133], each article with a different perspective on why. Both are interesting, the first focusing on practical timeseries analysis and the latter in general formulation and connection to Granger causality, which as a concept applies to stochastic timeseries. Schreiber [132] also introduces the thought process that leads to (7.2). See the paper by Ragwitz for the homonymous criterion [134] mentioned in Sect. 7.2. More about transfer entropy can be found in *Introduction to Transfer Entropy* by Bossomaier et al. [135] (but also in [76], Chap. 14). Also see *Directed information measures in neuroscience* by Wibral et al. [136] for how these ideas are routinely used in neuroscience and Faes et al. [355] with applications in physiology. More advanced methods for estimating mutual information and transfer entropy, that are mainly used in practice, are based on probability estimates using nearest neighbor searchers, see method of Kraskov et al. [356]. Note that transfer entropy can also be defined using distributions of ordinal patterns (i.e., permutation entropy) [137].

Regarding causal inference and dynamic influence, several methods exist. Krakovská et al. [138] performed a comprehensive comparison of many tests (and thus summarizes them as well). Another good reference is [139] by Runge et al., which discusses the problem of causal inference in complex systems and categorizes the status quo. The software CausalityTools.jl has implementations of most causal inference methods. The convergence cross mapping technique was introduced by Sugihara et al. in 2012 [140], and see also [141]. However, an early approach for estimating interrelation between chaotic variables which is very similar to the CCM method was suggested already in 1991 by Čenys et al. in [142]. Other nonlinear state space reconstruction based methods have been introduced by Arnhold et al. [143] and Hirata et al. [144].

Regarding timeseries surrogates, the work that introduced the concept first is by Theiler et al. in 1992 [145]. A detailed review paper with several applications and pitfalls was recently published by Lancaster et al. in [146] (and an older review in [147]), while a comparison of the discriminatory power of popular discriminatory statistics was done by Schreiber in [148]. The example we presented here is from Lancaster et al. [146]. A software to create surrogates is TimeseriesSurrogates.jl.

## Selected Exercises

7.1 Create two timeseries $x = \cos(t)$, $y = \cos(t + \phi)$ for some given $t$. As a function of $\phi$ calculate their Pearson coefficient, the Spearman coefficient and their mutual information (with significance given by $\mu \pm 2\sigma$). For which of the three measures does the truth-ness of the statement "$x$ and $y$ are correlated" depend on $\phi$?

7.2 Load dataset 11. Using the mutual information and the process described in Sect. 7.1, show that these two timeseries are *not* correlated. *Hint: you might want to de-trend the data first.*

7.3 Calculate mutual information using the statistics of ordinal patterns (used to define permutation entropy, see Sect. 6.4.3). The probabilities $P$ in (7.1) are now the probabilities of each ordinal pattern for a given order $d$. The joint probabilities are calculated using the frequencies of co-occurrence of ordinal patterns of each of the timeseries. Apply this new MI method to the timeseries $x$, $y$ coming from a chaotic Rössler system, (4.1).

7.4 Continue the above exercise. For a given $d$ calculate the relative MI, which is the value of the MI of the real data divided by $\mu + 3\sigma$ where $\mu$, $\sigma$ describe the parameters of the null PDF as in Fig. 7.1. How does the relative MI depend on $d$? Repeat this for the timeseries of dataset 11.

7.5 Clearly, all three variables of the Lorenz-63 system, (1.5) are dynamically connected. Calculate the transfer entropy between all 6 possible pairs of variables and report your results. Which pairings have higher transfer entropy? Can you make sense of this from the rule $f$?

7.6 Repeat the above exercise for various values of the box size parameter that is used to calculate the corresponding probabilities. Are your results robust?

7.7 Load exercise datasets 14, 15 (each having 2 columns). For each dataset, calculate the transfer entropy between its components. Can you use the results to claim whether one of the two timeseries is a dynamic driver of the other? *Hint: as the timescale of influence is not known, compute the TE for several forward-time delays and average the result.*

7.8 Continue the above exercise: how do your results depend on the choice of $d_x, \tau_x, d_y, \tau_y$ used in the definition of TE, (7.2)? And what about the bin sizes you use to do the computation?

7.9 Simulate $x, y, z$ timeseries for the Lorenz system, (1.5) for the default parameters and sampling time $\Delta t = 0.1$. For all 6 possible pairs, calculate the convergent cross mapping coefficient $r$ for $N$ from 100 to 10000 and plot all 6 curves. What do you notice for CCM values *from* the $z$ timeseries? Can you justify your observation? *Hint: Read carefully Sect. 6.1.2.*

7.10 Apply CCM to exercise dataset 16 using columns 1, 2 and 4, which are respectively eccentricity, climatic precession and insolation reaching the Earth (results of a simulation). Calculate CCM from $1 \rightarrow 4$, $2 \rightarrow 4$, $4 \rightarrow 1$ and $4 \rightarrow 2$. Plot these for various $N$ from 500 to the full data length. Reason about your results. *Hint: First plot the timeseries themselves and see how they connect with each other.*

7.11 Write your own code that creates FT and AAFT surrogates as explained in Sect. 7.4. Then replicate Fig. 7.5 without using TimeseriesSurrogates.jl and confirm your code is valid.

7.12 Show that either of the columns of dataset 11 are indistinguishable from a linear stochastic process with added trend. Because the data are short, a more suitable discriminatory statistic $q$ is the self-mutual-information, averaged over delays 1–6. *Hint: you should first de-trend the data and use the AAFT surrogates. Alternatively, you can use the Truncuated-Fourier-Transform surrogates [149], suitable for data with trends, and implemented in* TimeseriesSurrogates.jl.

7.13 Simulate the Lorenz-96 model, (5.10) for $D = 5$, $F = 8$ and keep the $x_1, x_2, x_3$ timeseries. Implement the Ragwitz criterion discussed in Sect. 7.2 (this requires you to already have implemented nonlinear timeseries prediction as in Sect. 6.4.1) and use it to find optimal $d, \tau$ for the timeseries $x_1, x_2, x_3$. How do these compare with other techniques you learned in Chap. 6? *Hint: calculate $n_c$ as in Exercise 6.11 for various values of $d, \tau$, and pick the pair that maximizes $n_c$.*

7.14 Continue from above and calculate all pairs of $TE_{i \rightarrow j}$ for $i, j = 1, 2, 3$ and $i \neq j$. Given (5.10) would you expect strong or weak symmetry in TE when exchanging $i \leftrightarrow j$? What do your numerics show? *Hint: the absolute value of TE depends on the boxsize $r$ used to bin the data and estimate probabilities. We recommend to compute these TEs for various $r \in [1, 5]$. If the relative amplitudes of $TE_{i,j}$ stay the same for different $r$, then there is asymmetry.*

7.15 (This exercise is best to tackle after you've gone through Chap. 9) Simulate the coupled Rössler systems of (9.9), with $a = b = 0.2$, $c = 5.7$, $\omega_x = 1.12$, $\omega_y = 1.08$ and four different values for

$$(k_x, k_y) \in \{(0.075, 0), (0.05, 0.05), (0.0375, 0.0375).$$

Keep only coordinates $x_1$, $y_1$ and plot them as in Fig. 9.11. Think how the transfer entropy $TE_{x \to y}$, $TE_{y \to x}$ between these two timeseries should be and then compute it and discuss accordingly. Use $(0.0, 0.2, 0., 0.11, 0.19, 0.1)$ as the initial condition.

7.16 Repeat the above exercise but now use the convergent cross mapping as a measure of dynamic influence.

# Chapter 8
# Billiards, Conservative Systems and Ergodicity

**Abstract** Dynamical billiards are a subclass of conservative dynamical systems, which are paradigmatic because they are simple, familiar with everyday life, yet still accommodate almost all dynamical features of conservative systems. In this chapter we use them as examples to illustrate the dynamics of conservative systems, highlighting the coexistence of chaotic and regular orbits in the state space. Ergodicity is a concept historically linked with billiards and fundamental for dynamical systems theory, so we discuss it in this chapter. Related with ergodicity are recurrences, and the many practical applications they have in classifying dynamical systems, which we also mention in this chapter.

## 8.1 Dynamical Billiards

A billiard is a simple dynamical system in which a point particle is propagating freely inside a domain $\Omega$, typically on a straight line,[1] until it encounters the boundary of $\Omega$, $\partial\Omega$. Upon reaching the boundary the particle is reflected specularly and elastically (conserving velocity magnitude), as illustrated in Fig. 8.1. This elasticity condition makes billiards conservative systems[2] (see Sect. 1.5 for the definition). Motion in a billiard is therefore separated into three variables $x$, $y$, $\phi$, with $x$, $y$ the coordinates and $\phi$ the propagation direction of the particle (because the velocity magnitude never changes, we only consider its orientation). $\phi$ is an angle variable (cyclic, taken with modulo $2\pi$).

---

[1] There are also billiard systems with trajectories not following straight lines but geodesics, like magnetic billiards and Riemannian billiards.

[2] For readers familiar with Hamiltonian mechanics, the potential function is $V(x, y) = 0$, if $(x, y) \in \Omega$ and $V(x, y) = \infty$ otherwise.

G. Datseris and U. Parlitz, *Nonlinear Dynamics*, Undergraduate Lecture Notes in Physics, https://doi.org/10.1007/978-3-030-91032-7_8

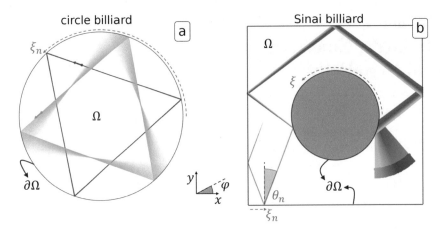

**Fig. 8.1** Two exemplary dynamical billiards: (a) the circle billiard, where all motion is regular, and (b) the Sinai billiard, where all motion is chaotic. In the circle billiard, a periodic (purple) and a quasiperiodic (cyan) trajectory are shown. In the Sinai billiard, a bundle of 100 initially parallel particles is evolved as shown in Code 8.1. Animations of this figure are available online under animations/8/circle_billiard and animations/8/sinai_billiard

Billiards are great systems to illustrate aspects of dynamical systems, because they feel naturally intuitive to us humans.[3] Probably the most relevant reason this happens is because light rays behave very much like particles in a billiard on human spatial scales (so-called "geometric optics"). But also in games like pool/snooker, or pinball, the ball behaves like a particle in a billiard. Many physics areas have used billiards for modelling, like electrons moving in nanodevices, light scattering, lasers, wave physics and acoustics, tokamaks, quantum chaos, among others. And besides all that, billiards display almost all of the possible dynamics of conservative systems, as we will see below.

### 8.1.1   Boundary Map

Another reason why billiards are useful in this education context is that they allow you to explore both a continuous as well as a discrete version of the same system, by considering their *boundary map* (also called *Birkhoff coordinates*). Because between collisions the evolution of the particle is in a sense trivial, one only cares about the collision points of the particle with the boundary $\partial\Omega$ of the billiard and what is the angle of the particle at those points. Let $\xi$ denote the position along the boundary of the billiard and $\theta$ the angle of the particle with a vector normal to the boundary

---

[3] That is why we used billiards to make a short educational video about deterministic chaos and the concepts of Chap. 3, which you can find online: youtube.com/watch?v=svV1MsUdInE.

at the collision point ($\theta$ is taken exactly *after* the collision). Then the variables $\xi_n$ and $\sin(\theta_n)$[4] define the variables of the boundary map of the billiard. Here $n$ is the collision number. The boundary map is a Poincaré map of the billiard, with the surface that defines the section being the actual boundary of the billiard! See Fig. 8.2 for a boundary map example.

## 8.1.2 Mean Collision Time

Billiards can have arbitrary and complex shapes. For all physical applications of billiards it is useful to know what is the average time between collisions $\tau$ (always starting from the boundary). This $\tau$ is also referred to as *mean free path*, a fundamental notion in electron transport or wave transport (free time and free space coincide in a billiard since velocity magnitude is always 1). So, how does one go about to calculate $\tau$?

Astonishing as it may be, one doesn't have to look far. It turns out that there is a universal and remarkably simple formula for it arising from basic geometric properties of the billiard

$$\tau = \frac{\pi |\Omega|}{|\partial \Omega|} \tag{8.1}$$

with $|\Omega|$ the area (that particles can be in) of the billiard and $|\partial \Omega|$ the length of the billiard boundary (total arclength $\xi$).

**Code 8.1** Simulating particles in the Sinai billiard using DynamicalBilliards.jl, resulting in Fig. 8.1b.

```julia
using DynamicalBilliards, PyPlot
bd = billiard_sinai()  # load pre-defined billiard
ax = plot(bd)          # plot it
# initialize a beam of 100 parallel particles:
x0, y0, φ0, N, dx = 0.2, 0.75, π + π/4 + 0.414, 100, 0.002
ps = particlebeam(x0, y0, φ0, N, dx)

for (i, p) in enumerate(ps) # iterate over particles
    x, y = timeseries(p, bd, 8) # evolve for 8 collisions
    color = (i/N, 0, 1 - i/N, 0.5)
    ax.plot(x, y; color) # plot trajectory
end
```

---

[4] Why $\sin(\theta)$? Consider a line segment of length $d\xi$. A parallel ray incident with angle $\theta$ on that segment "sees" length $\cos(\theta)d\xi$, i.e., the length element of the segment does not only depend on its actual length, but also on $\theta$. Using $\sin(\theta)$ instead of just $\theta$ normalizes the area element $d\xi\, d\theta$ of the boundary map because $d \sin(\theta)\, d\xi = \cos(\theta)d\theta\, d\xi$.

### 8.1.3   The Circle Billiard (Circle Map)

The circle map is the simplest discrete dynamical system that can display quasiperiodicity (Sect. 2.4), given by $a_{n+1} = (a_n + 2\pi\omega) \mod 2\pi$. If $\omega = m_1/m_2$ is rational, $a_n$ will undergo periodic motion with period $m_2$. However, if $\omega$ is irrational, the sequence $a_n$ will densely fill the unit circle and display quasiperiodic motion.

It turns out that the circle billiard is a system fully equivalent with the circle map, and not because of the coincidentally same name! Thus, the circle billiard provides a unique pedagogical view on quasiperiodicity because it shows what it "looks like" in both a continuous and a discrete system. But first of all, let's explain why these two circle systems are the same. Due to the simple geometry of the circle billiard, there is actually no dynamics in the $\theta$ variable of the boundary map. This is because specular reflection conserves $\theta$ and all chords of a circle with same length have same angle. Therefore $\theta_{n+1} = \theta_n$ (you could also say "period 1"). The dynamics in $\xi$ is also straightforward: at each collision a constant amount of arclength $\zeta(\theta)$ is added to it, $\xi_{n+1} = (\xi_n + \zeta(\theta)) \mod \ell$, with $\ell$ the total arclength of the billiard, here $\ell = 2\pi$. Basic Euclidean geometry (inscribed angle is half the intercepted arc) gives $\zeta(\theta) = \pi - 2\theta$. Therefore, the equations for $\xi_n$ and for $\alpha_n$ are equivalent!

Let's look at two characteristic examples. The purple orbit of Fig. 8.1a has $\theta_0 = \pi/6$ which means that $\xi$ will have a period of exactly 3, and thus the total motion is periodic. The orbit in real space is an equilateral triangle in this case. We now look at the cyan orbit. This has $\theta_0 = \pi(1/6 + \nu)$, with $\nu$ a small irrational number. As a result, the cyan orbit doesn't "close exactly" in some sense. It will continue to precede, giving the impression of an almost equilateral triangle rotating slowly clockwise (see figure).

## 8.2   Chaotic Conservative Systems

### 8.2.1   Chaotic Billiards

Chaos in billiards arises because of collisions with curved boundaries[5] and because of non-trivial geometries leading to different times between collisions for different points on the boundary. In Fig. 8.1b we show an initially parallel but tightly packed beam of particles evolving in the *Sinai billiard*, a system that played a foundational role for dynamical systems theory. The figure shows some initially close particles (representing a perturbation blob) that grow apart exponentially fast when evolved. Well, the exponential part is not clear from the figure, as one needs to show the timeseries, but hopefully by now we have earned your trust!

---

[5] This is at least the case for billiards where $\Omega$ has flat curvature (straight propagation), as the ones we discuss here. For billiards with other types of curvature, it is possible to have chaos without curved boundaries as is the case in, e.g., Hadamard billiards or magnetic billiards.

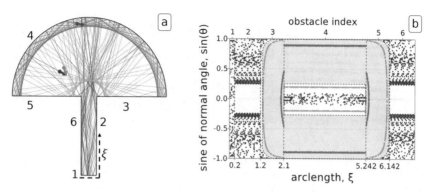

**Fig. 8.2** **a** The mushroom billiard, with one chaotic (purple) and two quasiperiodic (orange, cyan) orbits. **b** In the boundary map the regular part of the state space is enclosed by black dashed lines (and hued slightly grey). This part of the boundary map is an "island", isolated from the chaotic part (purple). The billiard is composed of several "obstacles" (walls and semicircles here, numbered from 1 to 6 as a guide to the eye)

What is really beautiful about chaos in billiards is how the mechanism of stretching and folding actually happen at the same place. The trajectories grow apart during a collision with a curved boundary (stretching in the $\phi$ coordinate). But at any collision with any boundary trajectories are naturally confined into the same bounded space $\Omega$ (folding), as they are blocked by the boundary! For periodic billiards the folding comes from the state space itself, and the same kind of folding happens for the $\phi$ coordinate in the long run, because $\phi$ is cyclic.

## 8.2.2 Mixed State Space

It is typically the case that conservative systems have a mixed state space. This simply means that orbits in some parts of the state space are chaotic, while some others are regular. Arguably the best example to demonstrate this is the mushroom billiard shown in Fig. 8.2. There quasiperiodic orbits exist that never go into the stem of the mushroom but instead stay in the cap.

Similar mixed state space structure became evident in Fig. 3.7 as well, where the purple-ish circles are quasiperiodic orbits (and can also be seen in the state space plot of the standard map in Fig. 8.3). The regular regions are dominated by quasiperiodic orbits and are often surrounded by chaotic regions. That is why sometimes the regular regions are called regular "islands", being inside a "sea" of chaos.

In these mixed state space systems, chaotic and regular state space regions are *isolated*. It is not possible for chaotic orbits to "enter" the regular regions and vice versa, no matter for how long we evolve them for. Furthermore, for most parameters, regular islands obey a self-similar structure, where near the "shore" of the island one finds smaller islands (corresponding to higher order periodic motion), and around the

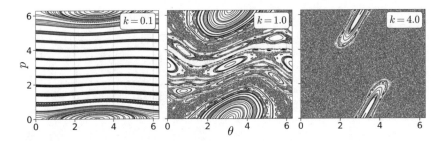

**Fig. 8.3** The standard map, (1.4), for three different values of the nonlinearity parameter $k$. A grid of initial conditions is evolved and scatter-plotted, and colored according to the maximum Lyapunov exponent (purple for positive, black for zero). Similar plot can be obtained from the mushroom billiard by tuning its parameter $w$, the width of its stem, from 0 to $2r$ with $r$ the radius of the cap

smaller islands one finds even smaller by continuing to zoom. A simple illustration of this can be seen online in animations/8/sm_zoom using the standard map as an example.

### 8.2.3  Conservative Route to Chaos: The Condensed Version

A conservative system without any nonlinearity (and thus no chaos) has a state space dominated by regular orbits (periodic and quasiperiodic). As an example in this section we will use the standard map, (1.4), which has a nonlinearity term $k \sin(\theta_n)$, tuned by $k$. We see in Fig. 8.3 that for small nonlinearity $k = 0.1$ the state space is still entirely regular.

As the conservative dynamical system becomes more strongly nonlinear[6] many regular orbits "break" (some become unstable while some stop existing). Chaotic orbits then take their place, and the remaining regular orbits form "islands" that are surrounded by the chaotic orbits, as seen in Fig. 8.3 for $k = 1$. As we make the system progressively more nonlinear, the "shore" of the regular islands becomes smaller and smaller, while smaller islands disappear completely ($k = 4$). What we've described here is a (very much) condensed version of the conservative route to chaos, whose beautiful theory is well outside the scope of this book, see Further reading for more.

---

[6] Several ways of increasing nonlinearity exist. E.g., by increasing a parameter that controls nonlinear terms or in nonlinear Hamiltonian systems by generating initial conditions of higher energy (typically).

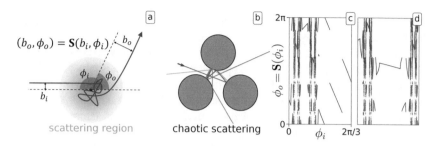

**Fig. 8.4 a** Illustration of the scattering function **S. b** Chaotic scattering in the 3-disk open billiard. Three particles with incoming angles differing only by 0.001 radians have massively different scattering. **c** Scattering function for the 3-disk open billiard. **d** A zoom-in of **c** between 0.1555, 0.1567 rad, establishing its fractal structure

### 8.2.4 Chaotic Scattering

Chaotic scattering is another form of *transient chaos* that can be presented intuitively by billiards. Besides the obvious example of billiards, another intuitive real-world scenario of chaotic scattering is with comets in gravitational fields of many bodies.

In a scattering problem a trajectory approaches a *scattering region* from infinity, gets scattered, and proceeds to leave towards infinity again. While outside the scattering region there is "no interesting dynamics", the dynamics within the scattering region can be highly chaotic. A scattering problem is formulated via the *scattering function* **S**, which connects incoming and outgoing states: $\mathbf{S}(\mathbf{x}_i) = \mathbf{x}_0$. In the example of Fig. 8.4, the incoming velocity of the particle doesn't matter, so the trajectory can be specified by two numbers: $\phi_i$, the incoming angle w.r.t. the scattering center, and $b_i$ the incoming offset. The scattering function outputs the same two numbers but now for the outgoing part: $(\phi_o, b_o) = \mathbf{S}(\phi_i, b_i)$, as illustrated in Fig. 8.4.

Perhaps the best example to illustrate chaotic scattering is the 3-disk open billiard of Fig. 8.4b. To simplify, we will set $b_i = 0$ drop $b_i, b_o$ from the equations, which implicitly assumes the size of the scattering region to be negligible. We then focus on estimating $\phi_o = \mathbf{S}(\phi_i)$, which we show in Fig. 8.4c. This function has *fractal* properties, due to its self-similarity, as is indicated by the zoom-in in Fig. 8.4d, which shows that the output of **S** has extreme sensitivity on the input.

## 8.3 Ergodicity and Invariant Density

Ergodicity is an important concept rooted deeply in the foundations of dynamical systems, stochastic processes, statistical physics, and the connection between these. Originally, ergodicity was discussed as a property of a dynamical system as a whole. Loosely speaking, ergodicity means that one has a "measure-preserving dynamical system" whose trajectories will visit all parts of the "available" state space. We've put these words in quotes, because, as will become apparent at the end of the section, what they mean really depends on how one defines things.

In general, however, in this book we want to move the discussion away from the system and into sets in the state space (e.g., discussing chaotic sets rather than chaotic systems). To this end, we will discuss a practically relevant definition of an *ergodic set*. An ergodic set is a subset $\mathcal{R}$ of the state space $\mathcal{S}$ of a dynamical system where for this set spatial averages (also called ensemble averages) and temporal averages coincide, i.e., $\mathcal{R}$ satisfies (8.2).

Imagine for a moment that you have some measurement function $g(\mathbf{x}(t))$ of the state $\mathbf{x}(t)$ and you want to know the average value of $g$ as a trajectory from a chosen initial value $\mathbf{x}_0$ evolves. Provided that $\mathbf{x}_0$ is part of an ergodic set, you could obtain this average by following a trajectory starting at $\mathbf{x}_0$ for infinite time and sampling $g(\mathbf{x}(t)$ at each time step, *or* by calculating $g(\mathbf{x}_0)$ for every $\mathbf{x}_0 \in \mathcal{R}$. This implicitly implies that the trajectory starting from $\mathbf{x}_0$ will come arbitrarily close to any state space point in $\mathcal{R}$. Putting ergodic sets into an equation we have

$$\lim_{T \to \infty} \frac{1}{T} \int_0^T g\left(\Phi^t(\mathbf{x}_0)\right) dt = \int_{\mathbf{x} \in \mathcal{R}} g(\mathbf{x}) \rho_{\mathcal{R}}(\mathbf{x}) d\mathbf{x}, \quad \forall \mathbf{x}_0 \in \mathcal{R} \qquad (8.2)$$

with $\Phi^t$ the flow of the dynamical system, see Chap. 1 (for discrete systems the time integral becomes a sum over discrete steps). Equation (8.2) is also called the *ergodic theorem* or *Birkhoff's ergodic theorem*.

The left part of the equation is straightforward, it is simply the time average of $g$ over $\mathbf{x}(t)$. Now the right hand side is the average over state space, but with a twist: a distribution $\rho_{\mathcal{R}}$ exists. This $\rho_{\mathcal{R}}$ is called the *natural* or *invariant density* of states in $\mathcal{R}$, normalized so that $\int_{\mathcal{R}} \rho_{\mathcal{R}} = 1$. Now, if it happens that $\mathcal{R} = \mathcal{S}$, i.e., all state space is a single connected ergodic set (modulo some subsets of measure 0), then the dynamical system itself is called ergodic. The Sinai billiard of Fig. 8.1 is an ergodic dynamical system, and it is in fact one of the first systems to be proven ergodic.[7]

What is crucial to understand is that indeed, for an ergodic set $\mathcal{R}$ every initial condition will visit an arbitrarily small neighborhood of every point of $\mathcal{R}$. But it is not necessary that every neighborhood must be visited with the same frequency! To connect with Chap. 5, $\rho_{\mathcal{R}}$ is in fact approximated by the histogram discussed in Sect. 5.2, obtained from an infinitely long trajectory (with $\mathbf{x}_0 \in \mathcal{R}$), and by also setting the limit of $\epsilon \to 0$. Equivalently, $\rho_{\mathcal{R}}$ can be computed by evolving a uniform distribution $\tilde{\rho}$ until it converges to $\rho_{\mathcal{R}}$. This convergence criterion is important, because $\rho_{\mathcal{R}}$ must be invariant, i.e., $\Phi^t(\rho_{\mathcal{R}}) = \rho_{\mathcal{R}} \, \forall t$. Each ergodic set $\mathcal{R}$ has its own invariant density $\rho_{\mathcal{R}}$ which is zero for $\mathbf{x} \notin \mathcal{R}$. Some example invariant densities are shown in Fig. 8.5.

The integral of $\rho_R$ is called the *natural* or *invariant measure* $\mu$

---

[7] See online animations/8/sinai_ergodicity to see how an initial condition in the Sinai billiard will slowly fill the entire state space.

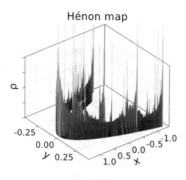
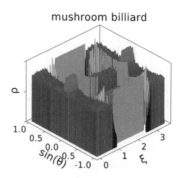

**Fig. 8.5** Approximate natural densities for the Hénon map chaotic attractor (3.3) and the boundary map of the mushroom billiard. The color and height both indicate the density value. For the mushroom billiard three densities are shown for the three trajectories of Fig. 8.2

$$\mu(\mathcal{B}) = \int_{\mathcal{B}} \rho_R(\mathbf{x}) \, d\mathbf{x}. \tag{8.3}$$

Assuming that $\rho_R$ is normalized, $\mu(\mathcal{B})$ is in fact the probability that an arbitrary state $\mathbf{x}$ (that is given to be in $\mathcal{R}$) will be in the subset $\mathcal{B} \subset \mathcal{R}$. Because $\rho_R$ is the differential of $\mu$, the following two syntaxes are equivalent: $\int \rho_R(\mathbf{x}) d\mathbf{x} \equiv \int d\mu$, and the latter is used commonly in the literature.

### 8.3.1 Some Practical Comments on Ergodicity

Being able to exchange ensemble averages with time averages has obvious practical benefit as depending on the application one might be more accessible than the other. However, there is another important consequence. Provided that you care about long term averaged properties in a chaotic ergodic set, then the pesky problem of sensitive dependence on initial conditions of Sect. 3.1 does not matter!

We have defined ergodic sets, but we haven't really discussed which sets are actually ergodic. In general, *attractive invariant sets* can be safely assumed to be ergodic. We also assume that every ergodic set has a invariant density $\rho$. Theoretically, to prove either of these things is exceptionally difficult. Now, because attractors have 0 volume in the state space, this makes it hard to identify the ergodic set with precision. That is also because the transient part of a trajectory, before convergence to the attractor, is *not* part of an ergodic set.

Specifically in conservative systems where trajectories stay bounded in state space the story is different. Given our practical definition of an ergodic set (i.e., any set that satisfies Eq. (8.2)), *almost all* state space points $\mathbf{x}_0$ belong to some ergodic set! You can easily identify this ergodic set by simply evolving a trajectory starting from $\mathbf{x}_0$ forwards in time. This means that (quasi-) periodic orbits in conservative systems can be considered ergodic sets, which is sometimes not the case in different definitions of ergodicity.

We can now finally return to the very beginning of this section where we mentioned the quoted "measure-preserving dynamical system". One would think that this measure must be the Lebesgue measure (the Euclidean volume), and the "available" state space to be all of state space. The Sinai billiard is a system that satisfies these criteria. But one can be much more flexible! An attractor $A$ is an invariant set and has an associated invariant measure $\mu_A = \int \rho_A$. If we consider our dynamical system only on the set $A$ of the attractor instead of the entire state space, and use $\mu_A$ as the measure, then this case satisfies the definition of an "ergodic system" since it "preserves the measure" $\mu_A$ and (by definition) visits all of the "available" space.

## 8.4 Recurrences

### 8.4.1 Poincaré Recurrence Theorem

The Poincaré recurrence theorem is a statement about bounded ergodic sets $\mathcal{R}$ of dynamical systems. Initially the theorem was stated specifically for conservative systems, but in fact it holds for any ergodic set with an invariant density, e.g., a chaotic attractor. The formal statement of the theorem is that for any open subset $\mathcal{B} \subset \mathcal{R}$ almost all trajectories starting in $\mathcal{B}$ will return there infinitely often. The *recurrence*, or *return time* is the time until a trajectory returns to the starting set for the first time. A consequence of the recurrence theorem is that any initial condition $\mathbf{x}_0 \in \mathcal{R}$, when evolved in time, will come arbitrarily close to $\mathbf{x}_0$ (i.e., revisit an $\varepsilon$-neighborhood around $\mathbf{x}_0$) infinitely often.

What is typically of interest is the statistics of the return times, because they can be used to characterize the ergodic set and to obtain dynamical information about it. Specifically, for orbits starting in $\mathcal{B}$ what is the average time $\langle t \rangle_B$ necessary to return to $\mathcal{B}$, and what is the distribution $p_B(t)$ of return times? Although there is no general system-independent formula, it is quite often the case that the return times are approximately power-law distributed, $p_B(t) \sim t^{-a}$ with some $a > 0$.

### 8.4.2 Kac's Lemma

Kac's lemma is a surprisingly simple formula that connects the average return time with the natural measure $\mu$ (or density $\rho$) of an ergodic set in the case of discrete dynamical systems. Kac's lemma states that the mean return iteration number $n$ to any subset $\mathcal{B}$ of an ergodic set $\mathcal{R}$ is given by the simple expression

$$\langle n \rangle_\mathcal{B} = \frac{\mu(\mathcal{R})}{\mu(\mathcal{B})}. \tag{8.4}$$

Thus, the smaller the measure of the subset $\mathcal{B}$, the longer it takes to return, which makes sense.

### 8.4.3 Recurrence Quantification Analysis

The statistics of recurrences in dynamical systems can be used to classify their behavior, quantify their properties and even estimate fractal dimensions among other things. The sum of numerical techniques that do this are typically referred to as *recurrence quantification analysis* (RQA), a nonlinear timeseries analysis technique. RQA is performed on the *recurrence matrix* R of a trajectory $X = \{\mathbf{x}_i, \ i = 1, \dots, N\}$

$$R_{ij}(\varepsilon) = \begin{cases} 1 & \text{if} \ \ ||\mathbf{x}_i - \mathbf{x}_j|| \leq \varepsilon \\ 0 & \text{else} \end{cases} \tag{8.5}$$

where $\varepsilon$ is a recurrence threshold (i.e., how close should a trajectory be to itself to count it as a recurrence). An example of R is shown for the trajectories of the mushroom billiard in Fig. 8.6. In the plot we also highlight the two most typical structures of R, the diagonal lines $\ell$ and the vertical white spaces $r$.

The orange plot in Fig. 8.6b is characteristic of quasiperiodic motion, as it is composed of infinitely-extending diagonal lines (where we of course do not see the infinite part due to finite time calculation), which have a regular, structured spacing in the vertical (or equivalently, horizontal) direction. And this is intuitively expected: diagonal lines indicate parts of a trajectory that are close to other parts, and when both parts are evolved forwards in time, they stay close to each other. For a quasiperiodic orbit, this will happen for ever. For a chaotic orbit on the other hand (like the purple plot in Fig. 8.6a), this does not happen forever due to the existence of (at least one) positive Lyapunov exponent. This is why the plot is dominated by much shorter diagonal lines (but they are also typically more dense). Therefore the diagonal lines have finite length $\ell$ (in the orange plot the diagonal lines also have finite length, but this is due to finite timeseries length).

The vertical distance between such diagonal lines is a recurrence time $r$ (also shown in the zoom of Fig. 8.6a). While the recurrence times are not constant for neither quasiperiodic or chaotic motion, it should be clear from the figure that their distribution is much more "irregular" for the chaotic trajectory than the quasiperiodic. The RQA framework puts these visual observations into mathematical formulas which can be calculated from the recurrence matrix R. Thus it allows one to make concrete statements about the average length of diagonal lines $\langle \ell \rangle$ or

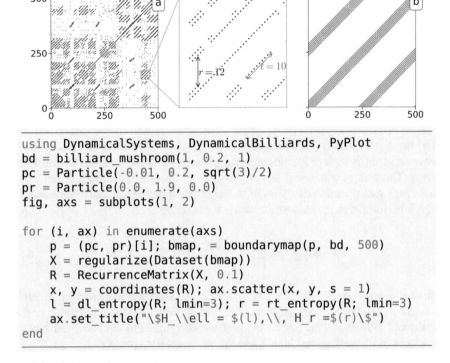

```
using DynamicalSystems, DynamicalBilliards, PyPlot
bd = billiard_mushroom(1, 0.2, 1)
pc = Particle(-0.01, 0.2, sqrt(3)/2)
pr = Particle(0.0, 1.9, 0.0)
fig, axs = subplots(1, 2)

for (i, ax) in enumerate(axs)
    p = (pc, pr)[i]; bmap, = boundarymap(p, bd, 500)
    X = regularize(Dataset(bmap))
    R = RecurrenceMatrix(X, 0.1)
    x, y = coordinates(R); ax.scatter(x, y, s = 1)
    l = dl_entropy(R; lmin=3); r = rt_entropy(R; lmin=3)
    ax.set_title("\$H_\\ell = $(l),\\, H_r =$(r)\$")
end
```

**Fig. 8.6** Recurrence matrices for the purple and orange trajectories (in boundary map representation) of the mushroom billiard in Fig. 8.2. The zoom-in defines some example diagonal line lengths $\ell$ and recurrence times $r$

the mean recurrence time $\langle r \rangle$,[8] or even their distributions, and see how these change with, e.g., the change in a parameter of the system, or across different experimental timeseries. For timeseries RQA is typically performed in the delay embedded space. Perhaps most useful for relative comparisons of trajectories (or timeseries) are the entropies (5.2) of diagonal line or return time distributions. In Fig. 8.6 we show values for both, $H_\ell$ and $H_r$, and it is clear that the chaotic timeseries has significantly larger entropies, as expected.

## Further Reading

Some good textbooks for chaos in conservative systems are *Chaos in Dynamical Systems* by Ott [28] and *Chaos and Integrability in Nonlinear Dynamics* by Tabor [150].

---

[8] $\langle r \rangle$ approximates the average of the average Poincaré recurrence time for all $\varepsilon$-sized subsets of the trajectory $X$.

KAM theory (from Kolmogorov, Arnold' and Moser, who all had influential role on dynamical systems theory) is one of the two major ingredients for understanding the conservative route to chaos, and is discussed therein. It shows how regular orbits "break" with the increase of nonlinearity. The Poincaré–Birkhoff theorem (also not discussed here), is the second necessary ingredient to understand the conservative route to chaos, along with the KAM theory, because it discusses what happens once a regular orbit "breaks". The ergodic theorem is most often attributed to Birkhoff, who also had foundational role for dynamical systems theory. Sinai, whom the billiard of Fig. 8.1b is named after [151], also had significant influence on understanding dynamical systems and billiards, and along with Kolmogorov played an important role in connecting deterministic dynamical systems with probabilistic concepts.

The most fitting textbook reference for this chapter is *Chaotic Dynamics: An Introduction Based on Classical Mechanics* by Tel and Gruiz [77]. Like the two previous books it has a lot of depth into theoretical concepts surrounding chaos in conservative systems. However, it also has several sections specifically about billiards as well as chaotic scattering. Natural measures, ergodicity, and their role in conservative chaos are discussed a lot by Tel and Gruiz, but also in the book by Schuster and Just [69].

Specifically for billiards, Chernov [152] contains a proof of the mean collision time in billiards, (8.1), and generalization to arbitrary dimensions. A recent work [68] connected mean collision times, Kac's lemma and Lyapunov exponents in billiards. Various types of billiards are described in [153], where sources regarding billiard applications in real world systems are cited therein, and see [154] for Hadamard billiards. Mushroom billiards were introduced in an influential work by Bunimovich [155] (and there it is shown explicitly how orbits staying in the cap of the mushroom are regular).

The basis of recurrence quantification analysis (RQA) is the Poincaré recurrence theorem [33]. The idea of RQA was initially stated by Eckmann, Kamphorst and Ruelle in [156] and then soon became a standardized tool in nonlinear timeseries analysis for data across several disciplines. A textbook on recurrence quantification analysis has been written by Webber and Marwan [157] but perhaps a better first read is the review paper by Marwan et al. [158], which has extensive practical focus and appropriate historic references. In [158] more quantities that can be calculated from a recurrence matrix are defined and used. Reference [159] contains information regarding practical usage of RQA and potential pitfalls. See also [160] for a recent application of RQA in billiards.

The best references that can serve as reviews for return time properties are probably the following two articles by Meiss [161, 162], who played an influential role in developing the theory of transport in conservative systems. Reference [161] also includes a proof of Kac's lemma. Specifically for the Poincaré recurrence theorem, one may read [33, 163, 164]. Interestingly, the scaling of the mean return time to an $\varepsilon$-ball of an ergodic set versus $\varepsilon$ can be used to estimate the fractal dimension of the set, as illustrated in [82] by Theiler.

**Fig. 8.7** A billiard with two
connected compartments

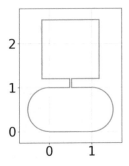

## Selected Exercises

8.1 Prove that a periodic trajectory of a discrete dynamical system is an ergodic
set (i.e., prove that (8.2) holds for it). What is the natural density of a periodic
orbit? What are the natural density $\rho$ and the measure $\mu$ of the state space of
a dynamical system that has a single, globally attractive fixed point at some
location $\mathbf{x}_0$?

8.2 Prove that the circle map $a_{n+1} = (a_n + v) \mod 1$ will densely fill the interval
$[0, 1)$ for irrational $v$. *Hint: show that the orbit $a_n$ will come arbitrarily close
to any point in the interval.*

8.3 Analytically calculate the mean return time to the disk of the Sinai billiard in
the boundary map formulation (i.e., the average number of collisions to return
to the disk) as a function of the disk radius $r$. Remember that the Sinai billiard
is ergodic. *Hint: calculate the natural measures of the disk and the whole state
space in boundary coordinates and use Kac's lemma.*

8.4 Create your own billiard simulation for a particle in the Sinai billiard. Use it to
numerically confirm the analytic result of the previous exercise. Also compare
with DynamicalBilliards.jl to ensure your code is valid.

8.5 Consider the logistic map for $r = 4$, $x_{n+1} = f(x_n) = 4x_n(1 - x_n)$. Show, both
numerically and analytically, that for this system, the interval $(0, 1)$ is an ergodic
set with invariant density $\rho(x) = \left(\pi \sqrt{x(1-x)}\right)^{-1}$. *Hint: iterate the given
density $\rho$ via the logistic map rule, and show that it remains invariant.*

8.6 In the exercises of Chap. 3 you proved (or at least we hope you did!) that
for a 1D discrete dynamical system, its Lyapunov exponent is given by the
expression $\lambda(x_0) = \lim_{n \to \infty} \frac{1}{n} \sum_{i=1}^{n} \ln \left| \frac{df}{dx}(x_i) \right|$. Using the information of the
previous Exercise 8.5 and the ergodic theorem, analytically show that for the
logistic map at $r = 4$ we have $\lambda = \ln 2$.

8.7 Create an animation of how an originally uniform distribution $\tilde{\rho}$ converges to
the natural density of the Hénon map.

8.8 Using DynamicalBilliards.jl (or alternatives you may code), create a billiard
like the one in Fig. 8.7. Then, create several thousand particles inside. While
evolving the particles, keep track of the ratio of number of particles above

height $y = 1$ and below. How does this ratio behave over time? Systems like these are called dynamical Maxwell's demons.

8.9 The width of the stem of the mushroom billiard $w$ is a parameter of the system. For a given $w$, populate the state space of the mushroom billiard with initial conditions. For each one, decide whether it is chaotic or regular (use one of the several methods you have learned in this book, or use the function `lyapunovspectrum` of DynamicalBilliards.jl). Separate the state space into regular and chaotic components and calculate the fraction of the state space that is chaotic. Plot this fraction versus $w$ from 0 to $2r$ (with $r$ the radius of the mushroom cap).

8.10 Evolve a trajectory of length $N = 10,000$ in the Sinai billiard (boundary map formulation) and using recurrence quantification analysis, obtain the mean recurrence time, the entropy of recurrence times, and the mean diagonal line length. Average the results for several initial conditions. How do these quantities behave versus the recurrence threshold $\varepsilon$ and versus increasing the disk radius $r$ of the billiard?

8.11 Compute in delay embedding space the recurrence matrices of the chaotic Rössler attractor ((4.1), $a = 0.2, b = 0.2, c = 5.7$), a periodic orbit of the Rössler system ($a = 0.2, b = 0.2, c = 2.5$), and a sequence of random numbers. Vary the length of the timeseries, the recurrence threshold $\varepsilon$ and the embedding dimension and interpret the results.

8.12 Load dataset 16 and (after normalization) for each of its columns create a recurrence matrix *without* embedding the timeseries. Vary the recurrence threshold $\varepsilon$ from 0.1 to 0.2 and calculate the entropy of diagonal and vertical lines as in Fig. 8.6. Report on your results, and comment whether there is a $\varepsilon$ value you would find more appropriate.

8.13 Continue from the above exercises, but now properly embed each of the 4 columns of dataset 16, using the methods you learned in Chap. 6. For each embedded timeseries, calculate a recurrence matrix in the reconstruction space using the Euclidean distance. Comment on how the recurrence plots, as well as the values for the entropies you obtained, differ from the non-embedded case.

8.14 Simulate a modified Sinai billiard which has a square instead of a disk in its center. Simulate particles and estimate (with any method you see fit), if there is chaotic motion. Can you justify your results?

8.15 Re-create the scattering problem of the 3-disk open billiard. Calculate and plot the function $n(\theta)$, where $n$ is the number of collisions happening between the 3 disks before the trajectory exits completely the scattering region.

8.16 Revisit Exercise 3.17. There we were interested in distinguishing trajectories that lead to evolution on a conservative 2-torus or a chaotic attractor. Calculate the natural density of these two sets in high precision using the initial conditions of Exercise 3.17. For this system trace$(J_f(\mathbf{x})) = 2(y + z)$. Since the evolution of volumes is given by $\dot{v} = \text{trace}(J_f(\mathbf{x})) \cdot v$, then the average value of $(y + z)$ gives us the average volume dissipation rate. Compute this value by (i) averaging it along each of the two trajectories, or (ii) averaging over state space weighted by the natural density as in (8.2). Confirm that for the conservative case you get an average rate of 0 and for the dissipative of $<0$.

8.17 Consider a periodic trajectory $X$ of length $N$ in the context of recurrence quantification analysis. In the limit of $N \to \infty$, $\varepsilon \to 0$, what values do you expect to obtain for the mean diagonal line length $\ell$, the mean recurrence time $r$ and the entropies of diagonal line lengths $H_\ell$ and recurrence times $H_r$?

# Chapter 9
# Periodically Forced Oscillators and Synchronization

**Abstract** Most dynamical systems in nature and technology are not isolated but interact with other systems. This interaction may change and adjust their dynamics, resulting in different forms of resonances and synchronization that are the topic of this chapter. We begin with nonlinear systems that are not capable of self-sustained oscillations (like, e.g., a damped pendulum). For this kind of passive oscillators periodic driving typically leads to characteristic resonances, bifurcations and transitions to chaotic motion if the forcing amplitude is high enough. In contrast, for systems which oscillate already without external driving, synchronization of the internal oscillation and the periodic forcing may occur. But even chaotic systems, when coupled, may exhibit different kinds of synchronization, including chaotic phase synchronization or generalized synchronization.

## 9.1 Periodically Driven Passive Oscillators

The simplest example of a periodically driven dynamical system in physics and engineering is the sinusoidally forced damped harmonic oscillator $\ddot{x} + d\dot{x} + \omega_0^2 x = a \sin(\omega t)$. This linear system describes, for example, RLC-circuits in electronics or mass-spring configurations in mechanics. It can be solved analytically and its response to the driving is often summarized in amplitude and phase resonance curves that you can find in any mechanics textbook. This model has three features that make it analytically solvable but at the same time not appropriate for many real world applications: the restoring force[1] $\omega_0^2 x$ is linear, the damping $d\dot{x}$ is proportional to the velocity $\dot{x}$ and the forcing is sinusoidal.

In the following we will focus on the impact of a nonlinear restoring force $r(x)$ instead of $\omega_0^2 x$. An example is the gravitational force $(g/l) \sin(x)$ of the forced pendulum $\ddot{x} + d\dot{x} + (g/l) \sin(x) = a \sin(\omega t)$, where $x$ is the angle of elongation, $g$

---

[1] The notion of a *restoring force* is used here referring to mechanical systems, but of course, this term may have other meanings in other contexts.

© The Author(s), under exclusive license to Springer Nature Switzerland AG 2022      137
G. Datseris and U. Parlitz, *Nonlinear Dynamics*, Undergraduate Lecture Notes in Physics,
https://doi.org/10.1007/978-3-030-91032-7_9

**Fig. 9.1** Amplitude resonance curves of the Duffing oscillator (9.1) showing the maximum $x_{max}$ of $x(t)$ versus the driving frequency $\omega$ for $d = 0.1$ and various $a$. To highlight the hysteresis the simulation is first run with increasing $\omega$ and then with decreasing $\omega$

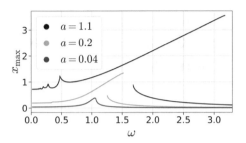

denotes the gravitational constant, and $l$ is the length of the pendulum. The restoring force $r(x) = x + x^3$ leads to the *Duffing oscillator*

$$\ddot{x} + d\dot{x} + x + x^3 = a\sin(\omega t) \tag{9.1}$$

that in the following will be used as a representative of nonlinear passive oscillators, that are not capable of self-sustained oscillations (such as the damped pendulum).

Figure 9.1 displays amplitude resonance curves of the Duffing oscillator (9.1) showing the maxima $x_{max}$ of $x(t)$ after transients decayed versus the driving frequency $\omega$. For small forcing amplitude $a = 0.04$ the curve resembles the resonance curve of the harmonic oscillator, because the nonlinear term $x^3$ is dominated by the linear term for small $x$. With increasing $a$, however, the maxima of the curves shift towards higher frequencies $\omega$ and for $a > 0.06$ two stable branches with very different amplitudes occur due to saddle-node bifurcations. These branches correspond to periodic solutions whose period equals the period of the driving $T = 2\pi/\omega$ and when following them by slowly varying $\omega$ we see a similar dynamical hysteresis phenomenon as with the 1D climate model (Fig. 4.1). In Fig. 9.1 this coexistence of two stable periodic oscillations with low and high amplitude occurs for $a = 0.2$ in a frequency range from $\omega \approx 1.25$ to $\omega \approx 1.5$. With an even stronger forcing amplitude $a = 1.1$ the hysteresis region is shifted towards higher frequencies and becomes wider. Furthermore, additional amplitude maxima occur for $\omega < 0.5$ constituting *nonlinear resonances* of the oscillator.

When increasing the driving amplitude $a$ furthermore, the impact of the nonlinearity $x^3$ becomes stronger and further bifurcations take place resulting in, e.g., period-$m$ orbits whose period is an integer multiple of $T = 2\pi/\omega$. At this point any illustration using only the maxima of $x(t)$ suffers from the fact that periodicity is not visible there. Therefore, from now on we use orbit diagrams based on the Poincaré map of the Duffing oscillator to illustrate and analyze the dynamics.

### 9.1.1  Stroboscopic Maps and Orbit Diagrams

Formally, the driven Duffing oscillator (9.1) can be written as a three dimensional autonomous system

**Fig. 9.2** Orbit diagram
(purple color) of the Duffing
oscillator for
$(a, d) = (7, 0.1)$. Insets
show examples of symmetric
(green) and asymmetric
(orange) orbits (in the
$x_1 - x_2$ plane)

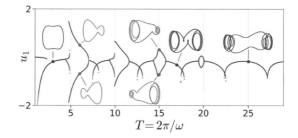

$$\dot{x}_1 = f_1(\mathbf{x}) = x_2$$
$$\dot{x}_2 = f_2(\mathbf{x}) = -x_1 - x_1^3 - dx_2 + a \sin(x_3) \qquad (9.2)$$
$$\dot{x}_3 = f_3(\mathbf{x}) = \omega$$

with $x_1 = x$, $x_2 = \dot{x}$ and $x_3 = \omega t \mod 2\pi$. The state space is the product $\mathbb{R}^2 \times S^1$ of the plane $\mathbb{R}^2$ spanned by $x_1$ and $x_2$ and the unit circle $S^1$ representing the dimension of the cyclic coordinate $x_3$. This topological structure of the state space suggests to introduce a Poincaré surface of section by specifying a value $c$ for the cyclic coordinate $x_3$ such that the section is taken at $\Sigma_c = \{(x_1, x_2, x_3) \in \mathbb{R}^2 \times S^1 : x_3 = c\} \cong \mathbb{R}^2$. Since $x_3$ represents the direction of time, it is guaranteed that trajectories will cross $\Sigma_c$ transversely and so this definition results in a well-defined Poincaré section. For any point $\mathbf{u} = (u_1, u_2) \in \Sigma_c$ the Poincaré map $\Pi(\mathbf{u})$ can be computed by using $(u_1, u_2, c)$ as initial condition of the Duffing system (9.2), integrating the ODEs over one period of the driving signal $T = 2\pi/\omega$ and then taking $\Pi(\mathbf{u}) = (x_1(T), x_2(T))$.

Practically this means that, if you want, for example, to plot a Poincaré section of a (chaotic) attractor of the Duffing oscillator you just solve the ODEs (9.2) and sample $x_1(t)$ and $x_2(t)$ with sampling time $T$. So you "switch on the light" periodically to observe $(x_1, x_2)$ at times $t_n = t_0 + nT$ and that's why in this case the Poincaré map is also called the *stroboscopic map*.

Having defined the Poincaré map we are now ready to compute orbit diagrams for the Duffing oscillator, just like we did in Sect. 4.2.2 for the Rössler system. Figure 9.2 shows an orbit diagram for damping constant $d = 0.1$ and driving amplitude $a = 7$ and some examples of symmetric and pairs of asymmetric orbits. Since resonances and corresponding bifurcations accumulate on the frequency axis at zero (see Fig. 9.1) more clearly arranged orbit diagrams are obtained with plotting versus the driving period $T = 2\pi/\omega$ instead of the driving frequency $\omega$.

There are many features one could analyze regarding the orbit diagram of Fig. 9.2. Here we will focus on the bubble-like structures that appear around periods $T_n = 5n, n \in \mathbb{N}$, for example in $14 < T < 16$, which resemble period-doubling bifurcations. However, the bifurcations taking place here are symmetry breaking bifurcations. The symmetry $r(x) = -r(-x)$ of the restoring force implies that $x_1$-$x_2$ projections of periodic orbits are either symmetric with respect to reflections at the origin or, if this symmetry is broken, occur as a pair of asymmetric orbits (see Fig. 9.2).

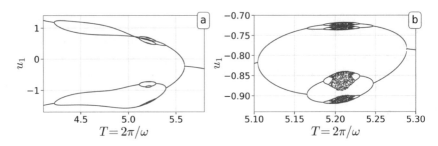

**Fig. 9.3** Orbit diagrams of the Duffing oscillator for $d = 0.1$ and **a** $a = 19.7$ and **b** $a = 19.75$ (notice that (**b**) is zoomed in)

Each branch at the symmetry bifurcation points in Fig. 9.2 thus corresponds to one of the two twin orbits which are generated simultaneously. They have the same periodicity and the same stability features, but you have to use different initial conditions to see both of them in the orbit diagram.

If the driving amplitude is increased asymmetric orbits undergo period doubling cascades[2] and reversed cascades that result in higher periodicity or chaos as shown in Fig. 9.3.

As can already be anticipated from Fig. 9.2 essentially the same bifurcation scenarios occur again and again when the driving period and the driving amplitude are increased. Such repeated bifurcation patterns are characteristic for periodically driven passive oscillators and occur also for other (coexisting) attractors, e.g., period-3, and their bifurcations (not shown here).

In systems with $\ddot{x} + d\dot{x} + r(x) = a\sin(\omega t)$ like the Duffing oscillator, saddle node, symmetry breaking and period doubling bifurcations are the only local bifurcations that can take place. Hopf bifurcations, invariant tori and quasi-periodic motion do not occur, because the Poincaré map $\Pi$ is everywhere contracting with $\det(J_\Pi) = \exp(-dT) < 1$. This is different with oscillators that exhibit self-sustained oscillations that will be considered in the next section.

## 9.2  Synchronization of Periodic Oscillations

### 9.2.1  Periodically Driven Self-Sustained Oscillators

Physical dynamical systems that possess an internal energy source may exhibit self-sustained oscillations (Sect. 2.2.2). A prominent self-sustained system is the *van der Pol oscillator* $\ddot{x} + d(x^2 - 1)\dot{x} + x = 0$. For small amplitudes $x$ the damping of this system (term multiplying $\dot{x}$) is effectively negative, leading to increasing $x$. This mechanism of instability is counter-balanced as soon as $x^2$ exceeds 1. Figure 9.4 shows typical oscillations for $d = 0.1$ and $d = 5$. For small values of the damping parameter $d$ the self-sustained oscillation is almost sinusoidal. Large $d$, however,

---

[2] Period-doubling bifurcations only occur for orbits with broken symmetry, so symmetry breaking has to precede period-doubling, see Exercises.

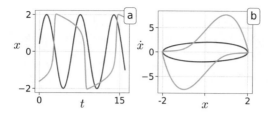

**Fig. 9.4   a** Timeseries $x(t)$ and **b** periodic orbits in the state space $(x, \dot{x})$ of the van der Pol oscillator for $d = 0.1$ (purple) and $d = 5$ (cyan)

result in so-called *relaxation oscillations* alternating between phases of fast and slow motion.

Figure 9.5a shows a Poincaré cross section (i.e., a stroboscopic phase portrait) of an attractor of the driven van der Pol oscillator

$$\ddot{x} + d(x^2 - 1)\dot{x} + x = a\sin(\omega t) \tag{9.3}$$

which has the same state space $\mathbb{R}^2 \times S^1$ as the driven Duffing oscillator and the same symmetry properties. The attractor in the surface of section is a subset of an invariant circle of the Poincaré map which corresponds to an invariant two-dimensional torus $T^2 = S^1 \times S^1$ in the full state space. The discrete dynamics on this circle can be described by an angle $\theta_n$ and Fig. 9.5b shows the resulting circle map $\theta_{n+1} = g(\theta_n)$.

With periodic forcing there is a competition between the internal (or natural) frequency of the free running oscillator ($a = 0$) and the driving frequency $\omega$. If both frequencies differ only a bit the van der Pol oscillator is entrained by the forcing, i.e., the resulting dynamics is periodic with a period $T = 2\pi/\omega$ and the van Pol oscillator is *synchronized* with the driving. Figure 9.6a shows an orbit diagram[3] where this kind of synchronization is visible as a window with period-1 attractors around $\omega = 0.5$.[4] If the driving frequency deviates too much from the natural frequency of the van der Pol oscillator synchronization with other frequency ratios occurs that leads to windows with period-$m$[5] attractors where $m > 1$. All these windows "open" and "close" with

**Fig. 9.5   a** Poincaré section of the van der Pol oscillator (9.3) with $(d, a, \omega) = (5, 1, 1)$ showing orbit points scattered on an attracting invariant circle. **b** Return map $\theta_n \to \theta_{n+1}$ of the angle characterizing the dynamics on the circle

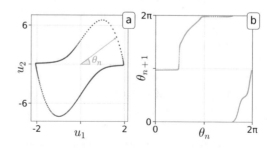

---

[3] Orbit diagrams of the forced van Pol oscillator are computed in exactly the same way as those for the Duffing oscillator.

[4] For $d = 5$ the period of the free running van der Pol oscillator is about $4\pi$, see Fig. 9.4.

[5] Period-$m$ for continuous systems means that the periodic trajectory crosses the Poincaré section $m$ times.

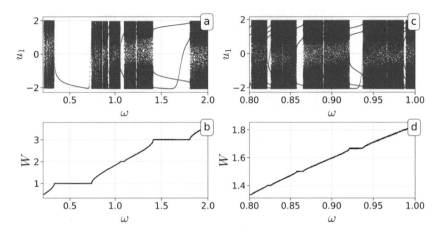

**Fig. 9.6** **a** Orbit diagram and **b** corresponding winding numbers of the van der Pol oscillator (9.3) for $d = 5$ and $a = 1$. (**c**), (**d**) Zoom of (**a**), (**b**)

saddle-node bifurcations. In some of them periodic oscillations with rather large periods occur represented by many points plotted on the vertical axis that must not be confused with chaotic solutions. The same holds for the quasiperiodic orbits which occur on the $\omega$ axis between the periodic windows. In fact, in the diagrams shown in Fig. 9.6 no chaos occurs (as can be checked by computing the largest Lyapunov exponent). Of course, if you increase the forcing amplitude beyond a certain threshold (see exercises), chaotic time evolution can also be observed.

The periodic and quasiperiodic solutions shown in Fig. 9.6 are all located on an attracting invariant torus in state space (see Fig. 2.5 for an illustration) whose Poincaré section is shown in Fig. 9.5a. Trajectories on this torus are characterized by two angles $\phi(t)$ and $\theta(t)$. The motion along the direction of time $x_3$ is given by $\phi(t) = \omega t$. The second angular motion takes place in the $x - \dot{x}$ plane of (9.3). It can be characterized by a protophase $\theta(t)$ that is obtained by computing the angle $\arctan(\dot{x}(t), x(t)) \in (-\pi, \pi]$ taking into account full rotations so that (on average) $|\theta(t)|$ increases proportional to time $t$. This motion can be described by a mean rotational frequency $\Omega = \lim_{t \to \infty} \theta(t)/t$, where the sign of $\Omega$ indicates clockwise ($\Omega < 0$) and counter-clockwise ($\Omega > 0$) rotations. Using both angular frequencies, $\omega$ and $\Omega$, we can define a *winding number* $W = \omega/|\Omega|$ and plot it versus $\omega$ as shown in Fig. 9.6b and in Code 9.1. In each period-$m$ window the winding number is constant and rational with $W = m/k$ where $m$ and $k$ are both integers. This appears as a "step" in the graph of $W(\omega)$, see Fig. 9.6b. Because there are infinitely many periodic windows, $W(\omega)$ has infinitely many steps and this is why it is also called a *devil's staircase*. If you zoom in you find periodic windows for any pair $(m, k)$ as illustrated in Fig. 9.6c, d.

Figure 9.7 shows regions in the $\omega$-$a$ parameter plane of the van der Pol oscillator with fixed rational winding numbers. These triangle shaped subsets of the parameter space where synchronization occurs are called *Arnold' tongues*. In the diagram, for clarity, only Arnold' tongues with $W \in \{\frac{1}{2}, \frac{2}{3}, 1, \frac{4}{3}, \frac{3}{2}, \frac{5}{3}, 2, \frac{7}{3}, \frac{5}{2}, \frac{8}{3}, 3, \frac{10}{3}, \frac{7}{2}\}$ are shown. In general, the larger the integers $m$ and $k$ representing the winding number $W = m/k$ of an Arnold' tongue the narrower the tongue and the smaller the "step"

**Fig. 9.7** $\omega$-$a$ plane of the van der Pol oscillator showing for $d = 5$ selected Arnold' tongues with winding numbers $W$ described in the text. Fig. 9.6b is a horizontal cut of this diagram at $a = 1$

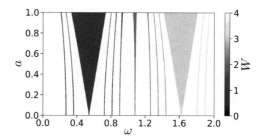

in the devil's staircase that you get when plotting the winding number as a function of $\omega$ for fixed $a$. Note that (impossible to resolve in a plot) each rational number on the $\omega$-axis at $a = 0$ is the tip of such a tongue with a triangular-like shape like those shown in Fig. 9.7.

**Code 9.1** Computation of the winding number of the forced van der Pol oscillator $W$ in the $\omega$-$\alpha$ parameter plane. Notice the lines that calculate the phase $\theta$ increment correctly, by taking into account the $2\pi$-warping.

```
using DynamicalSystems
function winding_number(ds, ω, a, u0)
    period = 2π/ω
    ds.p[2:3] .= (a, period) # set to new parameters
    tr = trajectory(ds, 4000, u0; Δt=period/20, Ttr=5000)
    tvec = 0:period/20:T
    u1, u2 = columns(tr)
    θ = 0.0; θ_old = atan(u2[1], u1[1])
    for (x, y) ∈ zip(u1, u2)
        θ_new = atan(y, x)
        θ += mod(θ_new - θ_old + π, 2π) - π
        θ_old = θ_new
    end
    W = ω/(abs(θ)/tvec[end])
    return tr[end], W
end

a_vec = 0:0.01:1; ω_vec = 0.2:0.02:2.0
ds = Systems.vanderpol(; μ = 5.0)
Wmat = zeros(length(ω_vec), length(a_vec))
u0 = ds.u0
for (ia, a) in enumerate(a_vec)
    a = a_vec[ia]
    for (iω, ω) in enumerate(ω_vec)
        u0, W = winding_number(ds, ω, a, u0)
        Wmat[iω,ia] = W
    end
end
```

## 9.2.2  The Adler Equation for Phase Differences

How typical or representative is the behavior of the van der Pol oscillator? To address
this question we recall that any periodic oscillation can be described completely
by a phase variable $\phi(t)$ parameterizing the motion on the closed limit cycle (see
discussion in Sect. 2.2.2) and that this phase increases with constant speed, i.e.,
$\dot{\phi} = \omega_0$. The dynamics of $\phi$ is neutrally stable, because perturbations of the phase
neither grow nor decay in time. Therefore, phases can be adjusted with relatively
small forcing. To investigate this we consider a periodically driven phase oscillator

$$\dot{\phi} = \omega_0 + \varepsilon Q(\phi, \omega t) \tag{9.4}$$

where $Q(\cdot, \cdot)$ is a driving function which is $2\pi$ periodic in both arguments. $\omega$ is
the forcing frequency and $\varepsilon$ the (small) forcing amplitude. Synchronization occurs
if for a given *detuning* $\Delta\omega = \omega_0 - \omega$ the phase difference $\Delta\phi = \phi - \omega t$ remains
bounded. This can be analyzed using a Fourier expansion of $Q$ and some averaging
over fast oscillations. This reasoning, that we will not present in detail here (see
Further reading), leads to a differential equation for the phase difference that is
called the *Adler equation*

$$\frac{d\Delta\phi}{dt} = \Delta\omega + \varepsilon \sin(\Delta\phi). \tag{9.5}$$

Synchronization with $\Delta\phi \to$ const. occurs if (9.5) possesses a stable fixed point.
This is the case if $|\Delta\omega| < \varepsilon$, i.e., within the Arnold' tongue shown in Fig. 9.8.

This analysis indicates that synchronization in Arnold' tongues is not a particular
feature of the driven van der Pol oscillator, but occurs with any periodically driven
system that exhibits periodic oscillations when running autonomously. And even
more, the same analysis can be done for a system of two bi-directionally coupled
phase oscillators without external forcing

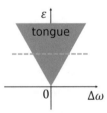

**Fig. 9.8** Arnold' tongue of the Adler equation (9.5). For a given $\varepsilon$, only values of $\Delta\omega$ that are
within the "tongue" lead to synchronization. The dashed line indicates a scan with fixed forcing
amplitude resulting in a plateau in the corresponding winding number diagram

$$\dot{\phi}_1 = \omega_1 + \varepsilon Q(\phi_1, \phi_2)$$
$$\dot{\phi}_2 = \omega_2 + \varepsilon Q(\phi_2, \phi_1). \tag{9.6}$$

Defining $\Delta\omega = \omega_1 - \omega_2$ yields for the phase difference $\Delta\phi = \phi_1 - \phi_2$ the same Adler equation (9.5) that now indicates when synchronization between two coupled periodic oscillators occurs.

### 9.2.3  Coupled Phase Oscillators

What happens if we have an ensemble of many, almost identical phase oscillators which are all coupled to each other? This is exactly the setting of the *Kuramoto model* of $N$ phase oscillators coupled all-to-all

$$\dot{\phi}_n = \omega_n + \frac{K}{N} \sum_{m=1}^{N} \sin(\phi_m - \phi_n). \tag{9.7}$$

The natural frequencies $\omega_n \in \mathbb{R}$ of the individual oscillators are assumed to be (slightly) different and in the following we will assume that they are Gaussian distributed $\omega_n \in \mathcal{N}(0, 1).$[6] $K$ denotes the coupling strength. The state of each oscillator can be represented by a point on an unit circle as shown in Fig. 9.9.

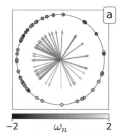
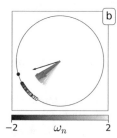
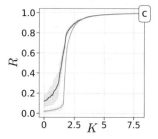

**Fig. 9.9** Phase dynamics of the Kuramoto model (9.7). **a** and **b** show phases $\phi_n(t)$ of $N = 50$ phase oscillators represented by points on a unit circle. The snapshots show the configurations at time $t = 200$ after starting the simulation with random initial conditions. Colors indicate the natural frequency $\omega_n$ of each phase oscillator. An orange arrow indicates the mean field. **a** For weak coupling $K = 1$ points are scattered on the circle while **b** with $K = 8$ all points form a cluster. **c** Mean value and standard deviation (shaded area) of order parameter $R$ versus coupling strength $K$ for $N = 50$ (purple) and $N = 2000$ (cyan) oscillators. An interactive application of this figure can be found online at animations/9/kuramoto

---

[6] The mean value of the natural frequencies is assumed to be zero, because this can always be achieved using a transformation to a coordinate system rotating with the mean velocity. The standard deviation can be set to 1 by rescaling time and $K$.

The color of each point indicates its natural frequency $\omega_n$. If the coupling is very weak the points move on the circle basically with their "individual speed" $\omega_n$ and the snapshot shown in Fig. 9.9a therefore shows a mixture of different colors. This changes if $K$ is increased and exceeds a critical value of $K_c \approx 2$. Now the oscillators form a narrow group led by the one with the highest natural frequency. While the others lag behind a bit they still can follow due to the coupling and perform a synchronous motion where all phase differences remain bounded, $|\phi_n(t) - \phi_m(t)| < \pi$. The transition to synchrony can be characterized by the complex mean field

$$R(t)e^{i\varphi(t)} = \frac{1}{N} \sum_{n=1}^{N} e^{i\phi_n(t)}. \tag{9.8}$$

If the phases $\phi_n$ are (almost uniformly) scattered on the circle then the colored arrows in Fig. 9.9a representing the complex numbers $e^{i\phi_n}$ point into opposite directions and the magnitude $R(t)$ of the mean field (orange arrow) oscillates at a very small value which converges to zero if the number of oscillators $N$ goes to infinity. However, once the coupling constant $K$ exceeds the threshold for synchronization $K_c$ all arrows point in almost the same direction (see Fig. 9.9b) and $R(t)$ reaches its maximum value of 1. The amplitude $R$ of the mean field is therefore used as an *order parameter* characterizing the transition to synchronization as a function of the coupling strength $K$ as shown in Fig. 9.9c. The magenta and the cyan curve show the temporal mean values $\bar{R}$ versus $K$ for $N = 50$ and $N = 2000$ oscillators, respectively, and the shaded regions indicate the standard deviations of the fluctuating order parameters. The larger the number of oscillators the sharper the transition.

## 9.3   Synchronization of Chaotic Systems

So far we studied synchronization of systems whose natural state was periodic. But what about coupling chaotic systems? Can they also synchronize and perform their chaotic oscillations in synchrony with each other? One might expect that this not possible, because sensitive dependence on initial conditions would immediately destabilize a synchronous state. But it turns that there are in fact several ways coupled chaotic systems can exhibit synchronized dynamics.

### 9.3.1   Chaotic Phase Synchronization

To have a concrete example we will couple two chaotic Rössler oscillators (see, e.g., Fig. 4.7). The full equations of the 6-dimensional system are

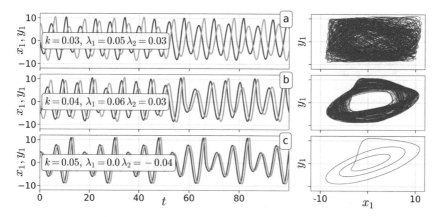

**Fig. 9.10** Coupled chaotic Rössler systems (9.9) with coupling $k_x = k_y = k$, $\omega_x = 0.98$ and $\omega_y = 1.02$. The legends show the values of $k$ and the two largest Lyapunov exponents (see also Fig. 9.12). From top to bottom the coupling increases and the dynamics transition from no synchronization to chaotic phase synchronization to periodic phase synchronization. An interactive application of this figure can be found online at animations/9/coupled_roessler.

$$
\begin{aligned}
\dot{x}_1 &= -\omega_x x_2 - x_3 & \dot{y}_1 &= -\omega_y y_2 - y_3 \\
\dot{x}_2 &= \omega_x x_1 + a x_2 + k_x(y_2 - x_2) & \dot{y}_2 &= \omega_y y_1 + a y_2 + k_y(x_2 - y_2) & (9.9)\\
\dot{x}_3 &= b + x_3(x_1 - c) & \dot{y}_3 &= b + y_3(y_1 - c)
\end{aligned}
$$

with $a = b = 0.2$, $c = 5.7$, leading to chaotic oscillations for the non-coupled case. Notice how we have modified the Rössler subsystems to have some innate frequency $\omega_x \neq \omega_y$, so that they are not identical (the case of identical coupled oscillators is rather special and is discussed in Sect. 10.2.1). Figure 9.10 shows simulations of the system for various values of the coupling strength.

For simplicity we use $k_x = k_y = k$ but the results we will see can occur also for $k_x \neq k_y$ if appropriate values are chosen. In any case, for weak coupling with $k = 0.03$ (Fig. 9.10a) there is no synchronization. Both subsystems oscillate with their natural frequencies $\omega_x$ and $\omega_y$, and the oscillations of the first subsystem (purple curve) are faster, because $\omega_x > \omega_y$. Since the oscillations of both subsystems are dominated by their natural frequencies, projections of the trajectories into the $x_1$-$x_2$ or $y_1$-$y_2$ plane exhibit a rotational motion that enables the definition of approximate protophases $\alpha_x, \alpha_y$ as illustrated in Fig. 9.11a. For very weak coupling the *phase difference* of the two subsystems increases in magnitude $|\Delta\alpha| = |\alpha_x - \alpha_y| \propto t$ due to the difference $\omega_x > \omega_y$ as shown in Fig. 9.11b.

Now if $k$ exceeds a critical value the oscillations occur "in phase", as seen in Fig. 9.10b. This is very similar to the transition we saw with coupled phase oscillators: phase differences are bounded if the coupling is strong enough. The only difference is that here the oscillations are not periodic but chaotic and therefore this state is called *chaotic phase synchronization*. Chaotic phase synchronization cannot be seen

**Fig. 9.11** **a** Projection of the
**x** chaotic Rössler subsystem
into the $x_1$-$x_2$ plane used to
define the protophase $\alpha$. **b**
Phase difference of two
coupled Rössler subsystems
(9.9) for coupling constants
$k = 0.03$ and $k = 0.04$

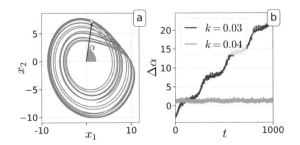

in all chaotic systems, but only in those which possess a strong periodic component
that enables the introduction of an approximate[7] phase.

Increasing the coupling even more results in periodic oscillations which are phase
synchronized (i.e., what we have seen in the previous sections) as shown in Fig. 9.10c.
There the motion in state space is truly periodic and can be uniquely described by
a phase (see Sect. 2.2.2). Let's take a moment to appreciate what's going on here.
The two subsystems are chaotic without the coupling. Naively one would expect that
coupling two chaotic systems would make the result even more chaotic. After all,
the state space dimension doubles and all the mechanisms that resulted in chaos in
the uncoupled case are still present. Yet we see that for strong enough coupling, the
motion ends up being periodic!

Okay, so far we have identified the three different cases by plotting the timeseries
and state space plots in Fig. 9.10. Separating the case of no synchronization from
phase synchronization (chaotic or not) can be done by computing the phase difference
$\Delta\alpha$ and checking whether the difference of the mean rotational frequencies $\Delta\Omega = \lim_{t\to\infty} \Delta\alpha(t)/t$ is zero, as shown in Fig. 9.12b. This, however, cannot distinguish
whether we have chaotic or periodic phase synchronization. Therefore, we show
in Fig. 9.12a the largest four Lyapunov exponents of the six-dimensional coupled
system (9.9) versus the coupling parameter $k$. Note that chaotic dynamics with one

**Fig. 9.12** **a** Four largest
Lyapunov exponents and **b**
frequency difference $\Delta\Omega$ of
the coupled chaotic Rössler
systems (9.9) versus
coupling constant
$k = k_x = k_y$ for
$(\omega_x, \omega_y) = (0.98, 1.02)$.
Note that coexisting
attractors with different
Lyapunov spectra may exist
for various parameter
combinations

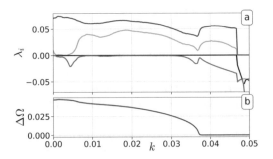

[7] The formal phase or the protophase (Sect. 2.2.2) uniquely parameterizes the motion on a closed
orbit. With aperiodic oscillations this is not a unique description anymore.

or even two positive Lyapunov exponents occurs without ($k < 0.037$: $\Delta\Omega > 0$) and with ($K > 0.037$: $\Delta\Omega = 0$) phase synchronization.

### 9.3.2 Generalized Synchronization of Uni-Directionally Coupled Systems

The two coupled Rössler systems in the previous section were different but not that much different. We will now discuss the case where two systems are coupled that have nothing in common and in general may even have different state space dimensions. Formally, for the uni-directional case this reads

$$\dot{\mathbf{x}} = f(\mathbf{x}), \quad \dot{\mathbf{y}} = g(\mathbf{y}, \mathbf{x}) \tag{9.10}$$

and our concrete example will be a chaotic Rössler system that drives a Lorenz-63 system, via its $x_2$ variable and coupling strength $k$

$$
\begin{aligned}
p\dot{x}_1 &= -x_2 - x_3 & \dot{y}_1 &= 10(-y_1 + y_2) \\
p\dot{x}_2 &= x_1 + 0.2x_2 & \dot{y}_2 &= 28y_1 - y_2 - y_1 y_3 + kx_2 \\
p\dot{x}_3 &= 0.2 + x_3(x_1 - 5.7) & \dot{y}_3 &= y_1 y_2 - 2.666 y_3.
\end{aligned}
\tag{9.11}
$$

Notice that we have used the parameter $p$ (with value 1/6 in the following), to adjust the timescale of the Rössler system to be closer to that of the Lorenz system.

In the generic case of uni-directional coupling, the concept of phase synchronization of the previous sections may not apply, as the forced system may not be appropriate to define an approximate phase. Instead, synchronization manifests by a *reproducible* response of the forced system to the driving system. Reproducible means that if you run this coupled system with some initial conditions for both subsystems, and you then repeat the simulation with the same initial conditions[8] for the driving system but different ones for the response system, you should still see the same oscillations of the Lorenz system after some transients decayed. This rather unexpected synchronization scenario is called *generalized synchronization*, and of course it only happens for sufficiently strong coupling (in our example for $k > k_{GS} \approx 6.66$). An interactive demonstration can be seen online at animations/9/generalized_sync.

Instead of repeating the entire coupled simulation, you can use the $x_2(t)$ variable of the Rössler system to simultaneously drive two identical copies of the Lorenz system $\dot{\mathbf{y}} = g(\mathbf{y}, x_2)$ and $\dot{\mathbf{z}} = g(\mathbf{z}, x_2)$ starting from different initial conditions $\mathbf{y}_0 \neq \mathbf{z}_0$. With generalized synchronization, their chaotic response will be asymptotically the same, i.e., $\lim_{t\to\infty} \|\mathbf{y}(t) - \mathbf{z}(t)\| = 0$. This test of generalized synchronization is known as *auxiliary system method*. Figure 9.13 shows the mean synchronization

---

[8] For this it is crucial that the solution of the driving system is exactly the same when repeating the simulation, and not affected by any perturbations, e.g., due to different step size control of the ODE solver.

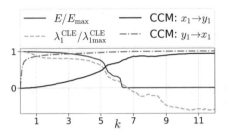

**Fig. 9.13** Generalized synchronization of a Lorenz system driven by a chaotic Rössler system (9.11). Plotted versus coupling strength $k$ are the synchronization error $E$ of the auxiliary system method, the largest conditional Lyapunov exponent $\lambda_1^{\text{CLE}}$, and the convergent cross mapping (Sect. 7.3.1) based on timeseries $x_1$ and $y_1$ from both systems. All quantities are normalized to a maximum of 1

error $E = \frac{1}{N} \sum_{n=m+1}^{m+N} \|\mathbf{y}(t_n) - \mathbf{z}(t_n)\|$ (after $m$ transient steps) of the two Lorenz copies versus $k$. $E$ becomes 0 for $k > k_{GS} \approx 6.66$.

An alternative approach to check whether the synchronized state $\mathbf{y} = \mathbf{z}$ is stable are *conditional Lyapunov exponents* (CLEs). There, the evolution of an infinitesimally small perturbation $\mathbf{e}$ of the state $\mathbf{y}$ is considered with $\mathbf{z} = \mathbf{y} + \mathbf{e}$ that is given by

$$\dot{\mathbf{e}} = g(\mathbf{x}, \mathbf{z}) - g(\mathbf{x}, \mathbf{y}) = g(\mathbf{x}, \mathbf{y} + \mathbf{e}) - g(\mathbf{x}, \mathbf{y}) = J_{gy}(\mathbf{x}, \mathbf{y}) \cdot \mathbf{e} \qquad (9.12)$$

where $J_{gy}$ denotes the Jacobian matrix of $g$ containing derivatives with respect to $y$. Solving this equation simultaneously with (9.10) provides the largest conditional Lyapunov exponent $\lambda_1^{\text{CLE}}$ which is negative for stable synchronization.[9] The name *conditional* LE indicates that the values obtained depend on the evolution of the state $\mathbf{x}$ of the driving system.

The reproducible response of the driven system that comes with generalized synchronization establishes a functional relation between the states of the drive and the response system. If the largest conditional Lyapunov exponent is sufficiently negative then the state of the response system $\mathbf{y}$ is asymptotically given by a function[10] $h$ of the state $\mathbf{x}$ of the driving system, i.e., $\lim_{t \to \infty} \|\mathbf{y}(t) - h(\mathbf{x}(t))\| = 0$. The graph $h(\mathcal{X})$ of the function $h$, where $\mathcal{X}$ is the state space of the driving system, is called the *synchronization manifold* and negative CLEs means that this manifold is (on average) contracting in the region of state space where the (chaotic) attractor of the driving system $\dot{\mathbf{x}} = f(\mathbf{x})$ is located. In general, strong uniform contraction leads to smooth functions $h$ and thus smooth synchronization manifolds. If both coupled systems are exactly the same, *identical synchronization* may occur and in this case the synchronization manifold is a linear subspace $\{(\mathbf{x}, \mathbf{y}) : \mathbf{x} = \mathbf{y}\}$.

---

[9] The algorithm for computing CLEs is identical to that for ordinary Lyapunov exponents outlined in Sect. 3.2.2. Equation (9.12) can also be used to compute a full spectrum of $d$ CLEs where $d$ is the dimension of $\mathbf{y}$, by simply using $J_{gy}$ as the Jacobian matrix.

[10] This statement can be made mathematically more rigorous, see Further reading for more information.

To detect generalized synchronization from timeseries (i.e., when you can't calculate CLEs or cannot use the auxiliary system method) you can use any method that aims at quantifying functional relations. One such method is the convergent cross mapping (CCM) we introduced in Sect. 7.3.1. In Fig. 9.13 CCM results based on $x_1$ and $y_1$ timeseries from both systems (9.11) are shown for embedding dimension $d = 6$ and delay time $\tau_d = 0.1$. Prediction of the $x_1$ timeseries from the $y_1$ is possible already for very weak coupling as indicated by the fast increase of the CCM measure $y_1 \to x_1$. This is due to the fact that the $y$ subsystem is driven by $x_2$ and $y_1$ therefore contains information not only about the dynamics in the $y$ subsystem but also in the $x$ subsystem. $y_1$ is therefore a suitable observable for reconstructing the state $(\mathbf{x}, \mathbf{y})$ of the full system (see Chap. 6) including $x_1(t)$.

In contrast, $x_1$ is *not* influenced by the dynamics of the response system, i.e., it does not contain any immediate information about $y_1$ and therefore the CCM measure $x_1 \to y_1$ remains rather small for coupling values below $k_{GS}$. This changes, however, with generalized synchronization, because now the emerging function $\mathbf{y} = h(\mathbf{x})$ enables predictions of $y_1$ from $x_1$. Unfortunately, there is no sharp transition at the onset of synchronization. The reason is that the function $h$ is for weakly contracting dynamics of the response system (as indicated by $\lambda_1^{CLE}$ near zero) not very smooth and thus difficult to approximate for the CCM algorithm using a finite set of data points. However, as soon as the largest conditional Lyapunov exponent of the response system becomes more negative $x_1 \to y_1$ (also) converges to one. This means that for GS with sufficiently strong coupling (here, e.g., $k = 12$) we can use the CCM method (or any equivalent tool) to conclude that there is coupling between the timeseries, but we *cannot* use it to discriminate between the uni-directional or bi-directional case.

As a final remark, we want to point out that identical or generalized synchronization also occur in coupled discrete systems (see exercises). In contrast, (chaotic) phase synchronization is a phenomenon that can only be seen in continuous systems because it requires an (approximate) phase to exist.

## Further Reading

The Duffing oscillator [165] is named after Georg Duffing, a civil engineer who investigated in the beginning of the 20th century the dynamics of power generators. Using a cubic restoring force and perturbation theory he showed the existence of two branches in the amplitude resonance curve [166] explaining the observed dynamical behavior. Like many other periodically driven passive nonlinear oscillators [167, 168] the driven Duffing oscillator possesses some specific pattern of bifurcations and coexisting attractors that repeats ad infinitum for decreasing driving frequency [169, 170]. These bifurcations and the corresponding periodic orbits can be characterized and classified by torsion numbers [171–173], linking numbers and relative rotation rates [174–176]. These concepts can also be applied to describe the topology of the flow in period doubling cascades [172, 177] and the topology of chaotic attractors

[178]. A physical example of a periodically driven system with nonlinear resonances are cavitation (gas) bubbles in a periodic sound field [179, 180].

The van der Pol oscillator was devised by Baltasar van der Pol [181] an electrical engineer and physicist who was working at Philips where he studied relaxation oscillations in electrical circuits with vacuum tubes. Synchronization with periodic driving was reported in his publication in 1927 with van der Mark [182]. There they also mention "Often an irregular noise is heard in the telephone receivers before the frequency jumps to the next lower values." This might be an early experimental observation of chaos. In fact, Cartwright and Littlewood proved in the 1940s that the periodically driven van der Pol oscillator may exhibit aperiodic oscillations [183]. Simulations with period-doubling cascades to chaos were reported by Parlitz and Lauterborn [184] and Mettin et al. [185] performed a detailed study of the van der Pol oscillator and its dynamics. A review on the work of van der Pol and the history of relaxation oscillations was published by Ginoux and Letellier [186].

Synchronization originates from the Greek and means "happening at the same time". It is a universal phenomenon occurring with neurons firing at the same time, male fire flies emitting light flashes together to attract females, or the applauding audiences that starts to clap hands in the same beat. The first scientific investigations of synchronization are attributed to Christiaan Huygens who studied in the 17th century the "odd kind of sympathy" of pendulum clocks hanging on the same beam. Later Lord Rayleigh observed synchrony of organ pipes. This and more historical notes on pioneering work on synchronization can be found in the book by Pikovsky, Rosenblum, and Kurths [187] which gives an excellent overview of synchronization phenomena and contains, e.g., also the derivation of the Adler equation. The popular book *Sync - How Order Emerges from Chaos in the Universe* by Strogatz [188] is an easy reading introduction. Another book on synchronization was published by Balanov et al. [189] and in the *Handbook of Chaos Control* [190, 191] you may also find several contributions on synchronization worth reading.

The Kuramoto model is named after Yoshiki Kuramoto who introduced it to study the dynamics and synchronization of large sets of weakly coupled, almost identical periodic oscillators [192]. It has been used as paradigmatic model for understanding many synchronization phenomena in diverse fields, including crowd synchrony on the London Millenium Bridge [193, 194]. A review of the extensive research on the Kuramoto model was published by Acebrón et al. [195].

Synchronization of chaos became popular in the beginning of the 1990s by the work of Pecora and Carroll [196] who demonstrated synchronization in unidirectionally coupled chaotic systems and proposed that this might be useful for private communication. A brief summary of work on chaos synchronization including references to earlier work in the 1980s is provided in [197] by the same authors and by Eroglu et al. [198]. These publications also address the phenomenon of *attractor bubbling*, an intermittent loss of synchronization due to (too) weak coupling, that was first observed and studied by Pikovsky and Grassberger [199] and Ashwin et al. [200, 201].

Chaotic phase synchronization was first reported by Rosenblum and Pikovsky [202] and has later been found in many systems including experiments, e.g., [203, 204].

Research on generalized synchronization began with work done by Afraimovich et al. [205]. Rulkov et al. [206] continued and coined the notion of *generalized synchronization*. The *auxiliary system method* was introduced by Abarbanel at al. [207]. Kocarev and Parlitz [208] discussed the emerging functional relation due to generalized synchronization and the role of subharmonic entrainment of chaotic systems [209]. Reference [210] contains a summary of different aspect of generalized synchronization and readers who are interested to learn more about its mathematical features are referred to publications of Stark [211] and Hunt et al. [212]. Generalized synchronization is exploited practically in machine learning algorithms based on reservoir computing (aka echo state networks [213, 214]). There the response system is a recurrent neural network and the nonlinear emergence of a dynamically induced nonlinear functional relation between input and output is used for predicting the future evolution of the input or some other quantities of interest [215, 216], including high dimensional spatiotemporal dynamics [217, 218].

## Selected Exercises

9.1 Show that if $\mathbf{x}(t) = (x_1(t), x_2(t), x_3(t))$ is a solution of (9.2) then $\tilde{\mathbf{x}}(t) = (-x_1(t), -x_2(t), x_3(t) + k\pi)$ is also a solution of this ODE system for any odd $k \in \mathbb{Z}$. Prove in the same way the equivalent statement that if $x(t)$ is a solution then $\tilde{x}(t) = -x(t + kT/2)$ is also a solution. If $\mathbf{x}(t)$ is a state on a periodic orbit then $\tilde{\mathbf{x}}(t)$ may either belong to the same orbit or another periodic orbit. If it is the same orbit the property $x(t) = -x(t + kT/2)$ holds and the projection of the orbit into the $x_1$-$x_2$ plane is symmetric with respect to reflections at the origin (see examples in Fig. 9.2). If $\tilde{\mathbf{x}}$ belongs to another orbit, this orbit is a mirror image of the first and thus a pair of asymmetric orbits exists. *Hint: To check whether a function $x(t)$ is a solution you should check whether the ODE is fulfilled.*

9.2 In Sect. 9.1.1 we mentioned that a symmetry breaking bifurcation must precede a period doubling bifurcation. Why is this the case? *Hint: perform a "proof by drawing". Can you draw a symmetric period-2 orbit?*

9.3 Show, based on the previous discussion of symmetry properties of the solutions of the Duffing equation that its Poincaré map $\Pi$ can be written as $\Pi(\mathbf{u}) = g(g(\mathbf{u}))$ with $g(\mathbf{u}) = -\Phi^{T/2}(u_1, u_2, c)$ where $\Phi$ is the flow, $\mathbf{u} = (u_1, u_2)$ are the coordinates in the Poincaré section and $c$ denotes the location of the section on the $x_3$ axis (see (9.2)). *Hint: try to explain, in words, the action of $\Pi(\mathbf{u}) = g(g(\mathbf{u}))$ step by step.*

9.4 Prove that for symmetric orbits with $x(t) = -x(t + kT/2)$ the Fourier spectrum of $x(t)$ consists only of odd multiples of the fundamental frequency. This is a very simple way to distinguish whether a given timeseries comes from a symmetric from a non-symmetric trajectory.

9.5 Compute a Poincaré section of the chaotic attractor of the Duffing oscillator at $(d, a, T = 2\pi/\omega) = (0.1, 19.75, 5.21)$ (see Fig. 9.3). Can the dynamics be described by a Lorenz map for one of the coordinates of the Poincaré map, like $u_1[n] \rightarrow u_1[n + k]$ with $n$ counting the iterations on the Poincaré map (analogous to the example shown in Fig. 4.7c)? *Hint: try different sizes of the iteration step $k$ and compare and interpret the results. For $k > 1$ consider subsequences $u_1[m], u_1[m + k], u_1[m + 2k], \ldots$ starting at different $m$.*

9.6 Find a period-3 attractor of the Duffing oscillator (9.1) for $(d, a, \omega) = (0.1, 12, 0.615)$, which coexists with other attractors, and compute orbit diagrams starting at this attractor for increasing and decreasing values of the control parameters $\omega$. What kinds of orbits and bifurcations do you find? *Hint: try different initial conditions to find the period-3 attractor.*

9.7 An example for a passive oscillator whose restoring force is not symmetric is the *Toda oscillator* $\ddot{x} + d\dot{x} + e^x - 1 = a \sin(\omega t)$. Compute orbit diagrams with control parameter $T = 2\pi/\omega$ for $d = 0.1$, $a = 0.88$ and $\omega \in (0, 2.2)$. Then increase the driving amplitude in small steps and repeat the computation. Compare the results with orbit diagrams of the Duffing oscillator as shown in Figs. 9.2 and 9.3. *Hint: in each step of the control parameter $T$ variation use the final state from the previous $T$ value for the new initial condition, both when increasing and decreasing $T$.*

9.8 Find a period-doubling cascade to chaos versus parameter $\omega$ of the van der Pol oscillator for $d = 5$ and $a = 5$. To do so compute orbit diagrams for periodic windows. *Hint: You may start your search in the period-4 window around $\omega = 2.45$. Then look what happens in other periodic windows nearby.*

9.9 Determine the fixed points of the Adler equation (9.5) and their stability.

9.10 The Poincaré map restricted to an attracting invariant circle, like Fig. 9.5b, represents a one dimensional $2\pi$-periodic circle map. Consider the prototypical *nonlinear circle map* $\theta_{n+1} = f(\theta_n)(\text{mod} 1)$ with $f(\theta_n) = \theta_n + \Omega - \frac{K}{2\pi} \sin(2\pi\theta_n)$. For this map the winding number is given by the mean number of rotations on the circle per iteration $W(K, \Omega) = \lim_{n\to\infty}(f^{(n)}(\theta_0) - \theta_0)/n$. Compute and plot the orbit diagram and $W(K, \Omega)$ versus $\Omega$ for $K = 0.1, 0.5, 1$.

9.11 Show that the Kuramoto model (9.7) can be rewritten as $\dot{\phi}_n = \omega_n + KR \sin(\varphi - \phi_n)$ where $R$ and $\varphi$ are the amplitude and the phase of the mean field (9.8), respectively. How would you interpret this result?

9.12 Repeat the analysis of Sect. 9.3.1, but instead of coupling Rössler systems, couple Colpitts oscillators, each given by

$$\dot{x} = \omega y - aF(z), \quad \dot{y} = c - \omega x - z - by, \quad \dot{z} = \epsilon(y - d)$$

with parameters $a = 30, b = 1, c = 20, d = 1, \epsilon = 1$ and $F(z) = \max(0, -1 - z)$. Couple two systems using the difference of the $y$ variables with coupling strength $k$, added to the equation of $\dot{y}$ (i.e., just like in (9.9)). For the two subsystems use $\omega_1 = 1.02$ and $\omega_2 = 0.98$. Identify $k$

values with no synchronization, chaotic phase synchronization, and phase synchronization. *Hint: how would you compute an approximate phase for these systems? The projection into the x-y plane will not work. Instead, produce projections into every possible plane, and decide which pair of variables is suitable to define an approximate phase, and what "origin" should be used, with respect to which you will define the phases.*

9.13 Consider the Lorenz-63 system, (1.5). For this system, the methods we used so far to define an approximate phase do not work, because the motion in, e.g., the $x$-$y$ plane performs an 8-shaped figure (and this is true even for different $\rho$ values where the motion is periodic instead of chaotic). How would you define a phase for such a system? You could create a new variable $u = \sqrt{x^2 + y^2}$ and plot this versus $z$. There it is clear how to define such a phase. Do this, and use the phase to calculate the mean rotational frequency $\Omega = \lim_{t\to\infty} \theta(t)/t$. Create a plot of $\Omega$ versus the parameter $\rho$ of the Lorenz-63 system.

9.14 Generate a chaotic signal $x_1(t)$ with a Lorenz-63 system $\dot{x}_1 = 10(x_2 - x_1)$, $\dot{x}_2 = rx_1 - x_2 - x_1x_3$, $\dot{x}_3 = x_1x_2 - (8/3)x_3$ for $r = 28$. Assume now that you don't know the value of $r$ and try find it by driving another Lorenz system $\dot{y}_1 = 10(y_2 - y_1)$, $\dot{y}_2 = rx_1 - y_2 - y_1y_3$, $\dot{y}_3 = y_1y_2 - (8/3)y_3$ with the signal $x_1(t)$. Determine $r$ by minimizing the averaged synchronization error $E(r) = \frac{1}{N}\sum_{n=1}^{N}(x_1(t_n) - y_1(t_n))^2$. *Hint: vary r in the driven system and compute the synchronization error as a function of r.*

9.15 Use the $x_1$ variable of the *baker map* $\mathbf{x}[n + 1] = f(\mathbf{x}[n])$ with

$$f_1(x_1, x_2) = \begin{cases} 0.5x_1 & \text{if } x_2 < c_1 \\ 0.5 + 0.5x_1 & \text{if } x_2 \geq c_1 \end{cases}$$

$$f_2(x_1, x_2) = \begin{cases} x_2/c_1 & \text{if } x_2 < c_1 \\ (x_2 - c_1)/c_2 & \text{if } x_2 \geq c_1 \end{cases}$$

to drive a one-dimensional discrete system $y[n + 1] = by[n] + \cos(2\pi x_1[n])$ with $c_1 = 0.49$ and $c_2 = 0.51$. Use the auxiliary system method and the largest conditional Lyapunov exponent to show that for $|b| < 1$ generalized synchronization occurs. Then plot $y[n]$ versus $x_1[n]$ to illustrate the resulting functional relation for $b = 0.3$, $0.7$ and $0.9$. In which sense do the graphs confirm statements about smoothness made in Sect. 9.3.2?

9.16 Show that with uni-directional coupling (like in (9.10)) the conditional Lyapunov exponents are a subset of the set of the Lyapunov spectrum of the full system. What do the remaining exponents in the spectrum represent? *Hint: Consider the block structure of the Jacobian matrix of the full system which implies similar block structures in the matrix Y representing the solution of the linearized equations and the matrices Q and R of its QR decomposition (see also Appendix A).*

# Chapter 10
# Dynamics on Networks, Power Grids, and Epidemics

**Abstract** Networks are one of the most commonly encountered structures in real life. They consist of nodes connected with each other, and can represent social networks, supply chains, power grids, epidemic spreading, and more. In this chapter we first introduce the basics of network theory and some of the major kind of networks one encounters. We then discuss two main topics typically modelled using networks: synchronization of many units and spreading of quantities. The first is applied to power grids and the second to spreading of epidemics. We also present agent based modelling, a framework that is used routinely in modelling epidemics, socio-economical systems, and other scientific disciplines.

## 10.1 Networks

Networks consist of units called *nodes* (also called vertices) that are connected by *links* (also called edges). These links may change in time, often on different timescales. For example, the brain is a typical example that is represented by networks. There the nodes are the neurons firing short pulses along their axons to other neurons. The axons and synapses provide the links between these nodes and their properties change in time relatively slowly due to learning, forgetting and aging. Other examples with almost static link structures are power grids (where the nodes are generators and consumers of electricity) or transportation networks with streets, railways, and flight routes.

### 10.1.1 Basics of Networks and Graph Theory

Mathematically a network can be represented by a graph $\mathcal{G}$. A graph $\mathcal{G} = (\mathcal{N}, \mathcal{L})$ is a pair of sets where $\mathcal{N}$ is the set of nodes and $\mathcal{L}$ the set of links connecting the

© The Author(s), under exclusive license to Springer Nature Switzerland AG 2022     157
G. Datseris and U. Parlitz, *Nonlinear Dynamics*, Undergraduate Lecture Notes in Physics,
https://doi.org/10.1007/978-3-030-91032-7_10

**Fig. 10.1** An undirected
(left) and directed (right)
network, along with their
adjacency matrices

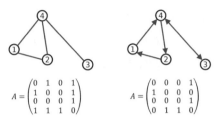

$$A = \begin{pmatrix} 0 & 1 & 0 & 1 \\ 1 & 0 & 0 & 1 \\ 0 & 0 & 0 & 1 \\ 1 & 1 & 1 & 0 \end{pmatrix} \qquad A = \begin{pmatrix} 0 & 0 & 0 & 1 \\ 1 & 0 & 0 & 0 \\ 0 & 0 & 0 & 1 \\ 0 & 1 & 1 & 0 \end{pmatrix}$$

nodes. We consider in the following finite networks with finite numbers of nodes
and links. Links between nodes can be *directed* of *undirected*. A directed link means
that some node has an impact on some other node but not vice versa or that an object
can move along this link in one direction only. In graphical illustrations of networks
directed links are represented by arrows (see Fig. 10.1). In contrast, undirected links
represented by a line only indicate that both nodes connected by them interact or
have something in common without making a difference for directions. If a network
consists of undirected links, only, it is called an *undirected* network. The structure
of a network, also called its topology, is fully described by its *adjacency matrix*
$A$ whose elements $A_{ij}$ are 1 if there is a link from node $i$ to node $j$ and 0 if not.
Adjacency matrices of an undirected network are symmetric, $A = A^{tr}$. One can also
use a *weighted* adjacency matrix $W$, whose non-zero elements have different values,
and could represent cases where some connections are stronger than others.

   An important concept to characterize a graph is the *node degree* $k_i = \sum_{j=1}^{N} A_{ij}$,
which is the number of edges connected to a given node $i$. A network whose nodes all
have the same degree is called *homogeneous*. If the nodes of a graph have different
degrees then the *node degree distribution*, i.e., the relative frequency of different node
degrees in the network, is an important statistical characterization of the graph. Many
other quantifiers exist for characterizing networks. A notable example is the *average
path length* of all shortest paths connecting any two nodes in the network. Another is
the *clustering coefficient* measuring the tendency of the nodes of a network to cluster
together. Clustering is defined as the percentage of the neighbors of a node that are
also neighbors with each other. Given an adjacency matrix it can be calculated as
$C_i = 1/(k_i(k_i - 1)) \sum_j \sum_\xi A_{ij} A_{j\xi} A_{\xi i}$ with $k_i$ the node degree as defined above.
Typically one is interested in the average clustering coefficient, averaged over all $i$.

### 10.1.2   Typical Network Architectures

With the term *network architecture* we loosely and qualitatively describe the connectivity patterns and the kind of network we have at hands. Some architectures are
typical in the sense of being ubiquitous in nature, being well-studied, and also having
clear characteristic properties, e.g., a specific node degree distribution. We display
some of them in Fig. 10.2 and discuss them further in this subsection.

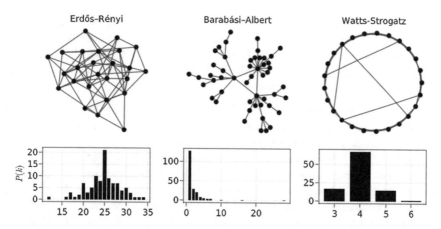

**Fig. 10.2** Examples of three typical networks architectures. Bottom row is node degree distribution (produced from larger networks for better accuracy)

**Random (Erdös-Rényi).** The simplest such architecture is that of *random networks*.[1] To construct such a network, one starts with a set of $n$ nodes and no links. Then, a link is established between each pair of nodes with probability $p$ (called the binomial model). As a result, their node degree distribution follows a Poisson distribution, i.e., most nodes have the same number of links, as one can show that the probability that a node has degree $k$ is $P(k_i = k) = B_{n-1}^k p^k (1 - p)^{n-1-k}$ with $B$ the binomial coefficient, giving a mean degree of $\langle k \rangle = pn$. Random networks also have small average path length, $\ell \approx \ln(n)/\ln(pn)$. Their clustering coefficient is $C = p$.

**Scale-free (Barabási-Albert).** Most real world networks do not have the aforementioned "random" architecture. Internet pages in the World Wide Web, for example, are linked in a way that is much different, because there are a few pages (e.g., search engines, journals, ...) with a very large node degree while many other pages, like private home pages, have only a few links. This results in a node degree distribution that follows approximately a power-law $P(k) \approx k^{-a}$ with constant $a \approx 2$ in case of the World Wide Web. From the illustration in Fig. 10.2 it becomes immediately clear that while for the random network there is a most likely node degree given by the maximum of the Poisson distribution, there is no such "typical" node degree with the power law distribution. Therefore, networks characterized by power-law distributions are called *scale free networks* and its few nodes with very high node degrees constitute the *hubs* of the network (which is why such networks are also called hub-worlds).

Many real world networks are scale free, due to the way they come into being. Starting with a small number $n_0$ of connected nodes, they grow by adding new nodes and links via the process of *preferential attachment*: new nodes have higher probability to be connected to nodes that already have more links. Specifically, each

---

[1] This name can be a bit misleading, as all network architectures we present can be created via a randomized process.

new node is connected to $n \leq n_0$ existing nodes with a probability that is proportional to the number of links that the existing nodes already have. The probability that the new node is connected to node $i$ is $P_i = k_i / \sum_j k_j$ with $k_i$ the node degree. In this way "the rich get richer" and hubs with very high node degrees occur. This generates heavy-tailed node degree distributions.

**Small-world (Watts-Strogatz).** Already several times in your life you may have thought "it's a small world!", for example, when you met a stranger, but it turned out that they know a person you also know. In fact it has been shown that on average with only six such friendship links you can connect to any other human on Earth (also known as *six degrees of separation*). This situation occurs when the network has a small average path length that scales logarithmically with the size of the network, which is the case for both random networks and scale-free networks. Here, we will distinguish a third kind of network, called *small-world* network, that in addition to small average path length also has unusually high clustering coefficients, independent of the network size. This property is observed in many real world networks like social networks.

You can generate a small world network starting with a ring network with $n$ nodes, each having $2K$ neighbors ($K$ on each side for each node). Then you start randomly "rewiring" the network. For each link of the network, replace it with probability $\beta$ to a different randomly chosen link (exclude self-connections and duplicate links in this process). The result is a network where approximately $\beta n K$ long-range links connect parts of the network that would otherwise be different neighborhoods. Even for very small $\beta$, the network already attains a very small average path length $\ell$ (compared to the unaltered network version which has high $\ell$), while the clustering coefficient remains very high even for large $\beta$.

### 10.1.3  Robustness of Networks

Most real world networks have some functional property: A power-grid delivers electricity, the neural network in the brain processes information, and in physiology the interaction of cells keeps an organ or organism running, to mention just a few. This raises the question of robustness of the network, or in other words, what happens if some node or link is not operating properly? Does such a defect have only a local impact in the network or will it lead to global failure of the functionality of the whole network? If the defect occurs accidentally, i.e., with the same probability for each node or link in the network, then scale-free networks are less vulnerable than random networks, because the vast majority of their nodes has only very few links and any damage thus remains localized without any major consequences for the full network. This robustness with respect to random defects could be a reason why natural evolution often led to scale-free networks in genetics or physiology, for example. The situation is different, however, with targeted attacks, where "important" nodes or links are first identified and then manipulated. In this case scale-free networks are

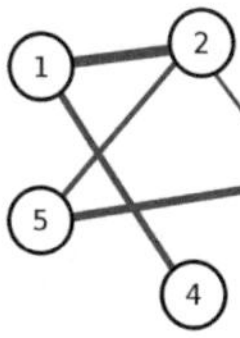

$$W = \begin{pmatrix} 0 & 7 & 0 & 5 & 0 \\ 7 & 0 & 3 & 0 & 4 \\ 0 & 3 & 0 & 0 & 6 \\ 5 & 0 & 0 & 0 & 0 \\ 0 & 4 & 6 & 0 & 0 \end{pmatrix}$$

$$\dot{x}_i = -y_i - z_i + \sigma \sum_{j=1}^{5} W_{ij}(x_j - x_i)$$
$$\dot{y}_i = x_i + a y_i$$
$$\dot{z}_i = b + z_i(x_i + c)$$

**Fig. 10.3** Network of five identical Rössler systems coupled in the $x$ ODEs by differences of their $x$ variables. The weighted adjacency matrix $W$ describes the mutual coupling between nodes which is controlled by the parameter $\sigma$

more vulnerable than random networks, because any attack on a node with a very high degree, i.e., a hub, may significantly affect the network's functionality.

## 10.2 Synchronization in Networks of Oscillators

In Chap. 9 we studied synchronization of pairs of dynamical systems. But of course, there are many examples where not only two systems are coupled, but a network of $N$ interacting systems exists.

To be more concrete Fig. 10.3 shows an example consisting of five coupled Rössler systems. To account for different mutual interactions a symmetric but weighted adjacency matrix $W$ is used to describe the coupling network. In addition, a global coupling parameter $\sigma$ controls the strength of all coupling links. In this example the coupling is provided by the $x$ variables and the coupling term $\sigma \sum_{j=1}^{5} W_{ij}(x_j - x_i)$ occurs in the first ODE of the Rössler system. But of course, one could also have this term in the second or third ODE for $\dot{y}$ and $\dot{z}$, respectively. And, instead of $x$ you may use $y$ or $z$ or some functions of the state variables to define the coupling. Which of these coupling options will lead to synchronization? And how does the answer to this question depend on the network structure given by $W$?

### 10.2.1 Networks of Identical Oscillators

It may seem hopeless to find satisfying answers to these rather general questions, but at least for networks of diffusively[2] coupled identical systems like the example shown in Fig. 10.3 there is hope! One can separate the impact of the network from the role of the local dynamics and the particular coupling using the so-called *master stability function*. With it we can answer questions like "Is there any network structure that will lead to synchronization for a given local dynamics?"

---

[2] Coupling based on differences of variables is often called *diffusive coupling*.

Identical, diffusively coupled dynamical systems can be generically expressed in the form

$$\dot{x}_i = g(x_i) + \sigma \sum_{j=1}^{N} W_{ij}[h(x_j) - h(x_i)] = g(x_i) - \sigma \sum_{j=1}^{N} G_{ij} h(x_j). \qquad (10.1)$$

Each subsystem is governed by a "local" dynamical rule $g$, has state $x_i \in \mathbb{R}^d$ and provides some output $h(x_i)$ that drives other nodes, depending on a weighted $(N, N)$ adjacency matrix $W$ with $W_{ii} = 0$. $\sigma$ is a global parameter for controlling the coupling strength and $G$ is a symmetric $(N, N)$ coupling matrix with $G_{ij} = -W_{ij}$ for $i \neq j$ and $G_{ii} = \sum_{j=1}^{N} W_{ij}$. By definition $G$ has zero row-sum ($\sum_{j=1}^{N} G_{ij} = 0$). The zero row-sum reveals that the state $x_1(t) = x_2(t) = \cdots = x_N(t) = s(t)$ is guaranteed to be a solution of (10.1) (this is the case here because the coupled dynamical systems are identical and with symmetric coupling). The situation where $x_i(t) = x_j(t) \,\forall\, i, j$ is a special case of synchronization called *identical synchronization* and is only possible when the coupled subsystems have exactly the same dynamic rule $g$. Furthermore, the condition $x_i(t) = x_j(t) \,\forall\, i, j$ defines a linear subspace in the full state space of the coupled dynamical system, constituting the *synchronization manifold*.

What we are interested to answer is the question whether this synchronization manifold is *stable* (i.e., attracting). It will probably come as no surprise that this can be done via linear stability analysis! However, in the current situation the linear stability analysis is a bit more extensive, see Appendix B for a derivation. The long story made short is that linear stability analysis of the synchronous state results in ODEs for $N$ perturbations $z_k(t) \in \mathbb{R}^d$ given by

$$\dot{z}_k = \left[ J_g(s) - \sigma \mu_k J_h(s) \right] z_k \qquad (10.2)$$

where $J_g$ and $J_h$ are the Jacobian matrices of $g$ and $h$, respectively, and $\mu_k$ $(k = 1, \ldots, N)$ are the eigenvalues of the coupling matrix $G$. The evolution of the synchronized state $s(t)$ is obtained by solving $\dot{s} = g(s)$ and (10.2) needs to be solved in parallel with $\dot{s} = g(s)$ (this is similar to what we did for computing the Lyapunov spectrum in Sect. 3.2.2). Note that except for the term $\sigma \mu_k$ all these linearized equations for the $N$ perturbations $z_k$ have the same form and their stability analysis can therefore by unified by considering the ODE system

$$\dot{z} = \left[ J_g(s) - \alpha J_h(s) \right] z \qquad (10.3)$$

where $\alpha$ is a control parameter that stands for $\sigma \mu_k$. Perturbations decay if the largest Lyapunov exponent $\lambda_1$ of (10.3) is negative. Since we want to asses the stability for different $\sigma \mu_k$ the value of $\lambda_1$ has to be known for a range of $\alpha$ values. This function $\lambda_1(\alpha)$ is called *master stability function*. It depends on $g$ and $h$, only, but *not* on the network structure given by the matrix $G$. This means that $\lambda_1(\alpha)$ can primarily tell us if there can be synchronization for *any* network structure, given the functions $h, g$ (i.e., is there a negative $\lambda_1$ at all?). Then, the network structure $G$ provides the actual

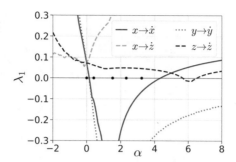

**Fig. 10.4** Master stability functions of the Rössler system with $(a, b, c) = (0.2, 0.2, 5.7)$ and different coupling terms. Black dots indicate the values of $\sigma \mu_k$ with $\sigma = 0.15$ for the network shown in Fig. 10.3. Except for $\mu_1 = 0$ all values lie in the range of $\alpha$ where $\lambda_{\max} < 0$ for $x \to \dot{x}$ and $y \to \dot{y}$, meaning that synchronization occurs for this network and these couplings

values of $\sigma \mu_k$ that $\alpha$ can obtain for a given network realization and tells us whether the given network can support synchronization (i.e., whether the values of $\sigma \mu_k$ fall in the range of $\lambda_1 < 0$). Examples of master stability functions for the Rössler system with different coupling terms are shown in Fig. 10.4.

Do we need stability $\lambda_1 < 0$ of (10.2) for all $k = 1, \ldots, N$? No, perturbations which grow but remain within the synchronization manifold are not relevant for stable synchronization. But how can we know which of the $k = 1, \ldots, N$ perturbations described by (10.2) grow transversally to the synchronization manifold and which stay inside? Fortunately, there is a simple answer. The zero-row-sum property of the coupling matrix $G$ implies that $G \cdot (1, 1, \ldots, 1) = 0$, i.e., $G$ has a vanishing eigenvalue $\mu_1$ with corresponding eigenvector $(1, 1, \ldots, 1) \in \mathbb{R}^N$. The first perturbation with $\mu_1 = 0$ corresponds to the perturbation dynamics $\dot{z}_1 = J_g(s)z_1$ *within* the synchronization manifold which may be governed by positive Lyapunov exponents in case of chaotic synchronization as in Sect. 9.3. For studying perturbations that may result in a loss of synchronization we therefore have to focus only on the stability of the other perturbations (10.2) with $k = 2, \ldots, N$. To do this you only have to insert $\sigma \mu_k$ for $\alpha$ in the master stability function and check whether $\lambda_{\max}(\sigma \mu_k) < 0$!

### 10.2.2 Chimera States

Loss of synchronization in networks of identical systems does not imply that *all* nodes get out of sync. It may happen that a subset of them remains synchronized while the rest shows incoherent dynamics. This is called a *chimera state*. Chimera states can be observed, e.g., in systems with some extended coupling range $R$. An example for such a configuration is a ring of $N$ phase oscillators which are coupled to $R$ nearest neighbors in both directions on the ring

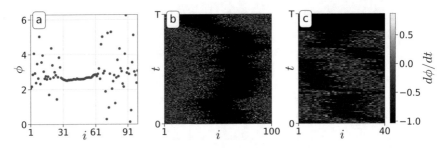

**Fig. 10.5   a** Snapshot of phases $\phi_i$ of network (10.4) with $N = 100$, $\omega = 0$, $R = 14$ and $\alpha = 1.46$ showing a chimera state. Nodes $i = 30 - 61$ are synchronized with almost equal phases while the phases of all other nodes are different. **b** Temporal evolution of the chimera state up to time $T = 25,000$. **c** Same as (**b**), but now for $N = 40$ phase oscillators. The chimera state collapses to a fully synchronous solution at $t \approx 22,000$

$$\frac{d\phi_i(t)}{dt} = \omega - \frac{1}{2R} \sum_{j=i-R}^{i+R} \sin[\phi_i(t) - \phi_j(t) + \alpha]. \tag{10.4}$$

The parameter $R$ specifies the range of the coupling and $\alpha$ is a phase lag that can be used to control the dynamics. Figure 10.5a shows a snapshot of the phases $\phi_i(t)$ where the synchronized phases constitute an almost horizontal line. Figure 10.5b shows a "space-time" plot where the synchronized part of the chimera state has dark blue color.

For sufficiently small networks (10.4) chimera states possess only a finite lifetime, i.e., after some period of time they suddenly collapse and a transition to the completely synchronized state takes place as illustrated in Fig. 10.5c. Such chimera states can thus be considered as chaotic transients, a phenomenon often observed in spatially extended systems (another example will be presented in Sect. 11.3.3).

### 10.2.3   Power Grids

A real world network of dynamical systems is the power grid. There, two types of nodes exist: generators which supply the grid with electrical power, and consumers of electricity. Since almost all power grids operate with alternating current at 50 or 60 Hz, synchronization of all generators is crucial for their functionality. Because we have different kinds of subsystems linked here (consumer vs. producer), we can't apply the theory of the master stability function. So we'll go with a trial-and-error approach (i.e., doing a simulation and checking for synchronization numerically). The state of each generator $i$ can be described by a variable $\theta_i(t) = \Omega t + \phi_i(t)$, where $\Omega$ equals 50 or 60 Hz. The network exhibits stable synchronization if the system converges to a state where all $\theta_i$ "run" with frequency $\Omega$. For this to occur, it is sufficient that all $\phi$ remain bounded, i.e., $|\phi_i(t)| \leq \pi$.

A detailed derivation of a model describing the evolution of the phases is beyond the scope of this presentation, however a simple model for a power grid network with $i = 1, \ldots, N$ units is

$$\ddot{\phi}_i = P_i - \gamma \dot{\phi}_i + \sum_{j=1}^{N} K_{ij} \sin(\phi_j - \phi_i). \qquad (10.5)$$

Here, $P_i$ stands for the electrical power which is either generated in node $i$ with $P_i > 0$ or consumed ($P_i < 0$). Parameter $\gamma$ is a damping constant representing dissipation of energy and $K$ is a weighted adjacency matrix describing the wiring topology of the network. The values $K_{ij}$ stand for the capacity of the transmission line between node $i$ and $j$. Equation (10.5) represents an extension of the Kuramoto model (9.7) with an inertial term $\ddot{\phi}_i$.

Now you may ask, what can we learn from such an idealized model about real existing power grids? As often in nonlinear dynamics, such a model may help understanding typical qualitative behavior. As an example, we will now study what happens if we increase the transmission capacity $K_{ij}$ along a link or add a link to an existing network. Naively one might expect that this will make the network "better" and thus support synchronous dynamics more easily. This, however, is *not* always the case! Even worse, increasing $K_{ij}$ or adding wires in the network may even result in a loss of synchronization, a phenomenon analogous to the *Braess's paradox* in traffic networks.[3] Figure 10.6 shows an example based on the dynamical model (10.5) on a small network of identical generators $P_i = P > 0$, identical consumers $P_i = -P$, and identical nonvanishing links $K_{ij} = K$. In the configuration shown in Fig. 10.5a all variables $\phi_i(t)$ converge to small constant values bounded by $\pm\pi$. If the transmission capacity between two nodes is locally increased by an amount of $\Delta K$ or if a new link is established then this may lead to a loss of synchronization in this network as illustrated in Fig. 10.6b, c.

Yet another idealization made to derive model (10.5) is the assumption that the power input and output at the nodes given by $P_i$ is constant. In real life this is, of course, not the case. Not only the demand by consumers varies in time but also power generation fluctuates, in particular with renewable energy sources relying on wind and sun. Abrupt changes in these parameters result in significant perturbations of the network that exceed the range of validity of any linear stability analysis. To cope with this challenge the concept of *basin stability* has been devised, estimating the probability that a given perturbation will kick the whole network out of the basin of attraction of the synchronized state. This approach is not only useful for studying the resilience of power grids, but also climate dynamics, and so it will be explained in more detail in Sect. 12.2.2.

---

[3] Adding more roads to a road network can sometimes slow down the overall traffic flow through it.

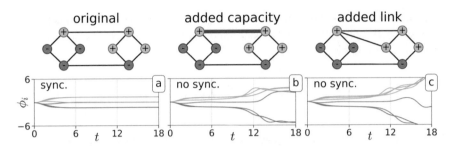

**Fig. 10.6** Realization of Braess's paradox in power grid networks using (10.5) and the network structure shown in the figure with consumers in orange and produces in cyan. Bottom plots show the time evolution of $\phi_i$. The original system (**a**) achieves synchronization. However, adding more capacity to the network, either via doubling the carrying load of one link (**b**) or adding an additional link (**c**), destroys synchronization

## 10.3   Epidemics on Networks

Search engines, social media, and machine learning enable unprecedented ways for manipulating opinion formation and propaganda. And only recently we learned the painful lesson of how a tiny virus (that shall not be named!) quickly spreads over the globe with very serious health and economic effects. What do opinion formation and epidemics have in common? In both cases personal interactions play a major role, and the spreading of opinions and viruses can formally be described using networks. In the following we will focus on the basic features of epidemic spreading, but many concepts used here can also be applied to opinion dynamics and related topics.

### 10.3.1   Compartmental Models for Well-Mixed Populations

Modeling the dynamics of infectious diseases began about a century ago with so-called compartmental models assuming a well-mixed population that can be divided into a few groups like: (S) *susceptible*, i.e., healthy individuals that are not immune and can be infected; (E) *exposed*, people who have been infected but are not infectious yet; (I) *infectious* individuals, who can infect others, and (R) individuals who *recovered* or died from the disease (also called *resistant*, because they are (temporarily) immune, or *removed* from the S and I compartments). Depending on the properties of the disease and the desired details some of these compartments may be neglected in the model or other compartments are added to account for special social groups, for example. Compartmental models describe the flux or transitions between these groups as illustrated in Fig. 10.7.

The simplest way to model such systems consists of deterministic ODEs providing the numbers of individuals in each compartment. Since dynamics of many infectious diseases is much faster than birth, death, and migration of individuals we will neglect these processes and assume that the total number of individuals in the population, $N = S\ (+E) + I\ (+R)$ is constant. The structure of the dynamical equations of

**Fig. 10.7** Illustration of the course of disease in three different compartmental models: SIS, SIR, SEIR. In the S(E)IR model recovered R are assumed to be immune forever while in the SIS model they become susceptible again. One can also take such a feature into account in extended versions denoted by S(E)IRS

**Fig. 10.8** Solution of the SIR-model (10.6) for $\beta = 0.3$, $\gamma = 0.1$ and $N = 10000$. Initial conditions are $(S, I, R) = (N - 1, 1, 0)$, i.e., 0.01% of the population are initially infected

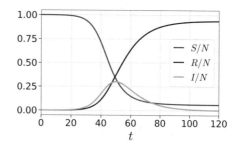

different compartmental models is similar. Therefore, we will consider now the SIR model (10.6) as a prototypical example and discuss its features in more detail. An infection may take place whenever a susceptible individual meets an infected/infectious person. The probability of such an encounter is for any susceptible given by $I/N$. Therefore the rate of infections is given by $\beta SI/N$ where $\beta$ is a parameter representing the infectivity of the particular virus and the number of contacts per time unit. While the number of infected $I$ increases with this rate, the number of susceptible $S$ decreases by the same amount. In addition to infection, there is also a recovery rate $\gamma$ that holds for all individuals in the same way. While the number of infected $I$ decreases due to recovery with a rate $\gamma I$ the number of recovered individuals $R$ increases with the same rate and we have

$$\dot{S} = -\beta \frac{SI}{N}, \quad \dot{I} = \beta \frac{SI}{N} - \gamma I, \quad \dot{R} = \gamma I. \tag{10.6}$$

Since the number $N$ of individuals is constant the number of recovered is given by $R = N - S - I$ and the dynamics of the SIR model (10.6) is completely described by the differential equations for $S$ and $I$.

Figure 10.8 shows an epidemic evolution of the SIR model (10.6) with a monotonously decreasing number of susceptible $S$, an increasing number of recovered individuals $R$ and a maximum of infected $I$ at the peak of the epidemic. Note, that while the number of infected goes to zero with $t \to \infty$ there remains a fraction of susceptibles that never got infected.

Whether the solution of the SIR model shows the typical course of an epidemic or not depends on the infection and recovery parameters $\beta$ and $\gamma$, respectively, and on the initial conditions $(S(0), I(0), R(0))$. An important quantity to assess the risk of an epidemic is the *basic reproduction ratio* $r_0$ that is defined as the number of new (also called secondary) infections caused by a single infected individual in a fully healthy but susceptible population with $S = N$, $I = 0$, $R = 0$. In this case there will be an epidemic if $\dot{I} > 0$ or, in the SIR model, $r_0 = \beta/\gamma > 1$.

The number of infections decreases if $\beta S/N - \gamma < 0$ or $S < \gamma N/\beta = N/r_0$. If the number of susceptible is below this threshold, e.g., due to previous infections and immunity or due to vaccinations, the disease cannot spread anymore and the population has reached the so-called *herd immunity*.

Another important quantity to characterize an epidemic is the total number of infected which equals in the case of the SIR model the asymptotic value $R_\infty = \lim_{t \to \infty} R(t)$, because finally all infected end up in the $R$ compartment. Last but not least the maximum number of infected $I_{max}$ is important because it might exceed the capacity of the heath care system.

The SIR model assumes uniform mixing across the entire population, i.e., any infected can infect any susceptible. In real life, however, social contacts are given by networks including clusters (family, school, workplace, . . . ) etc. This motivates to study the spread of an infectious disease on a network.

## 10.3.2   Agent Based Modelling of an Epidemic on a Network

There are truthfully a myriad of ways to model spreading of viruses. We already saw the basic approach of a three-dimensional ODE system (SIR model). Other approaches involve partial differential equations or stochastic modelling. An approach particularly suited to model epidemic spreading is *agent based modelling*, which also allows intuitively incorporating a network structure to the simulation.

Agent based modelling is a simulation where autonomous agents react to their environment and interact with each other given a predefined set of rules. These rules are not mathematical equations, as per the definition of a dynamical system, but instead formulated based on explicit statements (i.e., if condition X is fulfilled, do action Y). In a sense, agent based models are not traditional nonlinear dynamics, and are typically also not deterministic, but nevertheless they are universally used to simulate complex systems and emergent dynamics.

Let's set up a basic agent based simulation of the SIR model for virus spreading on a network. We start with a network (or graph). We populate each node of the network with a susceptible agent, and we let one agent start as infected. The agents have two properties (besides their identifier and position in the network): their status (susceptible, infected, recovered) and the number of days since infection. The rules of the simulation are then quite simple. All infected agents infect all neighboring[4] and susceptible agents, with some probability $\tau$. They then increment their number of days since infection, and if it crosses some threshold $p$ they recover. There are no rules for susceptible and recovered agents. In Fig. 10.9 we show the results of such a simulation, using the three introduced network architectures illustrated in Fig. 10.2. In Code. 10.1 we show how one can make such an agent based simulation. Can you guess and justify which network architecture would lead to faster virus spreading?

---

[4] Here neighboring means in the network sense: nodes connected with current node are neighbors.

**Fig. 10.9** Agent based
simulation of an SIR model
on three different networks:
Erdös-Rényi,
Barabási-Albert,
Watts-Strogatz. All networks
were made with 100 nodes
(and thus agents) and
approximately 200 links.
The curves show the mean $\pm$
standard deviation of 10
realizations of each kind

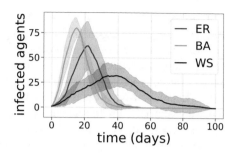

**Code 10.1** Agent based simulation of an SIR model on a network, using
Agents.jl and Graphs.jl.

```julia
using Agents, Graphs
@agent Human GraphAgent begin # Define the type of agent
    status::Symbol      # one of :S, :I, :R
    days_infected::Int
end
function init_sir_abm(g::AbstractGraph; τ = 0.1, p = 14)
    abm = ABM(Human, GraphSpace(g); properties=(τ=τ, p=p))
    # Add one agent to each node, with 1 infected agent
    starting_status(i) = i == 1 ? (:I, 1) : (:S, 0)
    fill_space!(abm, starting_status)
    return abm
end
# Rules of the ABM simulation (for each agent & each step)
function agent_step!(human, abm)
    human.status ≠ :I && return # not infected: do nothing
    # Infect all susceptible neighbors with probability τ
    for friend in nearby_agents(human, abm)
        if friend.status == :S && rand(abm.rng) < abm.τ
            friend.status = :I
        end
    end
    # Check if human recovers
    human.days_infected += 1
    if human.days_infected ≥ abm.p; human.status = :R; end
end
# run simulation and count infected and recovered
g = watts_strogatz(100, 4, 0.1)
abm = init_sir_abm(g)
infected(x) = count(i == :I for i in x)
recovered(x) = count(i == :R for i in x)
adata = [(:status, infected), (:status, recovered)]
adf, _ = run!(abm, agent_step!, 100; adata)
```

From here, the sky is the limit. One can make any number of adjustments to this model, e.g., add probability $d$ of dying that is checked for each day an agent is infected, or add a re-infection probability $\tau_r < \tau$ to re-infect a recovered agent, and so on. A more advanced scenario would be to incorporate transportation. So far, each node of the network is one agent. However, it is more realistic that each node of the network contains several agents (like a house, super market, ... ). The agents can then move between nodes with some probability, having higher probability to go towards a node that has many incoming links, i.e., a hub. In this scenario, agents would only attempt to infect other agents that reside in the same node, not neighboring ones.

An important modelling task is to study the impact of countermeasures and how this interplays with the network architecture. Once again, there are a lot of different ways one can model countermeasures: social distancing rules (reducing transportation probabilities or forbidding transportation to a given node), vaccines (reducing infection probabilities or out-right curing all agents at a specific node), among others. With agent based modelling doing such alternations to the simulation are straight-forward, and so we leave it for the exercises.

## Further Reading

Networks are structures that are used routinely in science and are ubiquitous in nature. By now, the topic has reached such a vast extent, that it is impossible to contain it in a single chapter. Worth having a look at are the highly-cited papers of Barabasi and Albert [219] and Watts and Strogatz [220], which sky-rocketed the field of network science. See also the subsequent [221] by Albert et al. on the robustness of networks.

As it is impossible to cite all relevant resources here, we point to the discussions and references within the following: a review article by Albert and Barabasi [222], a review article by Strogatz [223], another review by Newman [224], the exclusively online *Network Science Book* by Barabasi, the *Complex Networks* textbook by Latora [225] and the textbook by Chen, Wang and Li [226]. The books of Barabasi and Chen et al. have chapters on spreading on networks (e.g., epidemics). But also let's not forget the earliest analytical work on random networks by Erdös and Rényi, which stimulated many papers, see e.g., [227].

Reviews including nonlinear dynamics and synchronization on networks were published by Boccaletti et al. [228] and Arenas et al. [229]. Master stability functions can also be defined for more general cases. For example, if the coupling matrix is not symmetric but diagonalizable, then the dynamics of (10.3) has to be investigated for complex $\alpha$. Furthermore, extensions of the approach for networks consisting of almost identical systems and for networks of networks have been devised (but are beyond the scope of our presentation here).

The term "chimera states" refers to the monstrous fire-breathing hybrid creature in Greek mythology composed of the parts of three animals-a lion, a snake and a goat. It was coined by Abrams and Strogatz [230], but the coexistence of synchronous and asynchronous dynamics was first observed by Kuramoto and Battogtokh [231].

Wolfrum and Omel'chenko [232] pointed out the transient nature of chimera states in small networks.

Early studies of power grid dynamics using Kuramoto like models describing phase dynamics were presented by Filatrella et al. [233], Rohden et al. [234] and Dörfler et al. [235] to mention only a few. A collection of articles reflecting the current state of research on the dynamics of power grids is presented by Anvari et al. [236], see also the recent review by Witthaut et al. [357]. The Braess paradox is named after Braess who observed that adding a road to an existing road network can slow down the overall traffic flow [237]. Braess's paradox in oscillator networks was first studied by Witthaut and Timme [238] (which is where our Fig. 10.6 is inspired from). A Julia package that makes it simple to simulate continuous dynamics on a network is NetworkDynamics.jl.

The SIR model was introduced in 1927 by Kermack and McKendrick [12]. The most comprehensive treatment of epidemics on networks can be found in the book by Kiss et al. [239]. See also the network textbooks we cited above for chapters on epidemic spreading on networks.

Reference [240] by Bonabeau summarizes applications of agent based modelling on human systems, and see also the book by Hamill and Nigel [241] for applications in economics. Agents.jl is a Julia-based framework to perform agent based simulations. It integrates with Graphs.jl, which is a Julia library for graph creation and analysis and offers pre-defined functions to create the network types discussed in this chapters. An alternative well-known agent based modelling software is NetLogo.

## Selected Exercises

10.1 Using the algorithms discussed in Sect. 10.1.2, construct several random, scale-free and small-world networks. You can represent them via their adjacency matrix. Then, for each, calculate their node degree distribution writing your own version of a function that calculates it.

10.2 Generate scale free networks using growth and preferential attachment. Create different networks as follows. For each case start with a network of $n_0 = 5$ fully connected nodes, and grow it to one having $N = 1000$ nodes using preferential attachment with $n = 1, 2, 3, 4, 5$ respectively in each case. For each case, calculate the power-law coefficients of the node degree distribution. Does the coefficient increase, decrease, or stays constant with $n$ and why?

10.3 Generate random small-world networks for various $\beta$ and $K$ values. Calculate $\ell(\beta)$ and $C(\beta)$ and plot them versus $\log(\beta)$.

10.4 Implement the network of coupled Rössler systems shown in Fig. 10.3 and determine the range of $\sigma$ where synchronization occurs. Then replace the $x$ coupling in the $\dot{x}$ ODEs by an analog coupling in the $y$ ODEs using $y$ variables. Can you confirm the predictions of the master stability analysis (Fig. 10.4)?

10.5 Consider 10 Rössler systems located on a ring and coupled to their neighbors to the left and to the right, only. Compute the coupling matrix $G$ assuming that all mutual coupling strengths are equal. Does this network synchronize for the $x \rightarrow \dot{x}$ coupling also used in the example shown in Fig. 10.3? Use the master stability function shown in Fig. 10.4 to answer this question. If synchronization occurs, estimate the range of the coupling parameter $\sigma$ where this is the case. Now add an 11th Rössler system in the center of the ring which is connected to all other 10 oscillators in a star-shaped way. Use the same mutual connection strength. Are these additional links beneficial for synchronization? If so, in which sense?

10.6  Implement network (10.4) and estimate the duration of the transient chimera state for $N = 25, 30, 35, 40, 45, 50$. Use the order parameter

$$Z(t) = \left| \frac{1}{N} \sum_{k=1}^{N} e^{i\phi_k(t)} \right|$$

to monitor the synchronization state of the system and consider the time when $Z(t) > 0.95$ as the end of the chimera state.

10.7  How do synchronization and chimera dynamics of (10.4) depend on $\alpha$? Vary $\alpha$ around 1.46 and report your observations.

10.8  Implement the model of Fig. 10.6a. Define a function that computes the average synchronization error of the model by averaging the values of all $\phi_i$ and seeing whether this exceeds $\pm\pi$. Then, for each possible link in the model, increase the carrying capacity by $K$ (if there is no link it had 0 capacity and now it would have $K$, i.e., there would be a new additional link). Report in which cases synchronization is broken. Repeat the exercise by decreasing the capacity by $K$ (for links without an actual connection, i.e., zero capacity, no further decrease can happen).

10.9  Implement the SIR model and perform simulations for different values of $\beta$ and $\gamma$. How would you study the impact of a lockdown or a vaccination using this model? Try to simulate such scenarios. *Hint: which of $\beta$ and $\gamma$ is affected by a lockdown?.*

10.10  The SIS model is given by two ODEs: $\dot{S} = -\beta SI/N + \gamma I$ and $\dot{I} = \beta SI/N - \gamma I$. Show that $S(t) + I(t) = S(0) + I(0) = N = $ const.. Show that the SIS model can be reduced to a single ODE (for example) for $I$ and solve this ODE analytically using an ansatz $I(t) = c_1 \exp(c_2 t)/(c_3 + \exp(c_2 t))$. Use this solution to determine the asymptotic $(t \rightarrow \infty)$ values of $S(t)$ and $I(t)$ for both cases $\beta > \gamma$ and $\beta < \gamma$. For $\gamma = 0$ the SIS model turns into the SI model. How large is the percentage of (finally) infected in this model?

10.11  Use the first two ODEs of the SIR model to derive an expression for $dI/dS$ and solve this ODE analytically by separation of variables. Use the resulting relation between $I(t)$ and $S(t)$ to express $I$ as a function of $S$ and then use the condition $dI/dS = 0$ to determine the maximum number of infected $I_{max}$ as a function of $N, \beta, \gamma$ and the initial value $S(0)$.

10.12 Create your own agent based simulation (i.e., without using Agents.jl) of the simplest scenario discussed in Sect. 10.3.2 and shown in Code. 10.1. Confirm that you get the same results as in Fig. 10.9.

10.13 Implement the agent based model of SIR on a network with transportation as discussed in Sect. 10.3.2. Given a network, populate each of its nodes with $N$ susceptible agents. Make one agent infected. If an agent is infected for more than $p$ days, it becomes recovered. At each step of the simulation an agent may infect other agents in the same node with probability $\tau$. Then, the agent may change node with probability $m$. If transportation occurs, the agent moves randomly to one of the neighbor nodes, with probability directly proportional to the number of incoming links each neighboring node has. Produce a figure similar to Fig. 10.9 based on this simulation. *Hint: You can use Agents.jl and its* `move_agent!` *function to automate the transportation process.*

10.14 Extend the above simulation with the following extra rules: (i) recovered agents may be re-infected with probability $\tau_r < \tau$, (ii) if a node reaches a percentage $g$ of infected inhabitants, it becomes quarantined and transportation into and out of this node is not allowed for $q$ days. Come up with a measure of "total spread of virus" (single real number) and study how this number depends on the model parameters $\tau, \tau_r, p, g, q$. *Hint: these are many parameters. Keep 3 fixed and make a heatmap of the virus spreading number with varying the other 2. Repeat for a different choice of the 2 parameters that are not fixed.*

10.15 Agent based models (ABMs) are used routinely in ecological, social, or economical simulations. A famous example is Schelling's segregation model, which shows that complete segregation of individuals into groups with same identity can occur even if the individuals require less than 50% of their neighbors to have the same identity. Create an ABM simulation that displays this effect. Start with a two-dimensional square grid of finite size $M \times M$. Agents belong to one of two groups (A or B) and you populate the grid at random with $N < M^2$ agents (half from each group), so that each grid location can be occupied only by one agent. At each step of the simulation agents count their neighbors (maximum of 8). If they have at least $k$ neighbors of the same group, they are satisfied and do nothing. If they have less than $k$, they move to a random (but empty) location on the grid. Run your simulation and visualize the long-term state of the ABM. Report for which $k$ values you observe segregation. *Hint: if you can come up with a numerical quantifier of segregation, that's great! But simply plotting all agents with 1 color per group is easier.*

# Chapter 11
# Pattern Formation and Spatiotemporal Chaos

**Abstract** Many real world systems are extended in space and their state is given by functions of space and time describing fields like concentrations of chemical compounds, electric or fluid current densities or other relevant physical quantities. In such systems spatial patterns may emerge corresponding to spatial variations of the fields representing the states of the extended system. These patterns can be stationary or may oscillate in time resulting in periodic, quasi-periodic, or even chaotic spatiotemporal dynamics. As examples for pattern formation and (chaotic) spatiotemporal dynamics we consider the Turing instability in reaction-diffusion systems and spiral wave dynamics in excitable media that is, for example, underlying cardiac arrhythmias.

## 11.1 Spatiotemporal Systems and Pattern Formation

Most spatially extended systems in physics are described by *partial differential equations* (PDEs). In this case space and time variables are real numbers and the spatial domain is typically a connected subset of $\mathbb{R}$, $\mathbb{R}^2$, or $\mathbb{R}^3$. As a result, the dynamic variables of the system are no longer real numbers, but entire fields over the spatial domain. These evolve in time according to the PDE, hence the name *spatiotemporal systems*. Of course, any network of coupled neurons, oscillators or other dynamical units is in the real world also extended in space. There spatial proximity has an impact on coupling strength or coupling delays. Since these classes of coupled systems have been addressed in the previous chapters, here we will focus on PDE models.

### 11.1.1 Reaction Diffusion Systems

Many spatially extended systems display *pattern formation*. That is, their spatial

G. Datseris and U. Parlitz, *Nonlinear Dynamics*, Undergraduate Lecture Notes in Physics,
https://doi.org/10.1007/978-3-030-91032-7_11

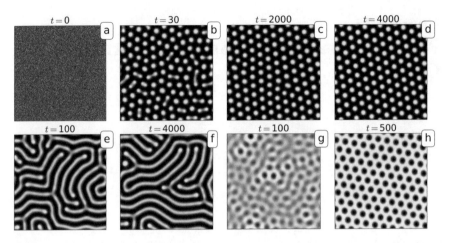

**Fig. 11.1** Snapshots of the $u_1$ variable of the Brusselator system (11.1) showing pattern formation for $a = 9$ and $b = 9.8$. All simulations start from random initial conditions as shown in **a**, which is white noise. **b–d**, **e–f** and **g–h** show snapshots of the formation of static pattern for $d = 4$, $d = 3$ and $d = 2$, respectively. Animations of the time evolution can be found online at animations/11/brusselator

fields will develop patterns over space, which may be stationary or oscillating in time (see Figs. 11.1 and 11.3 for examples). An important class of systems exhibiting pattern formation are reaction-diffusion models. They are used, for example, to understand morphogenesis, i.e., the formation of colored pattern in the skin or coat of animals.

A prototypical example of a reaction diffusion systems is the Brusselator (2.3) introduced in Sect. 2.2.2 but now extended in two spatial dimensions

$$\dot{u}_1 = 1 - (b + 1)u_1 + au_1^2 u_2 + d_1 \Delta u_1$$
$$\dot{u}_2 = bu_1 - au_1^2 u_2 + d_2 \Delta u_2 \tag{11.1}$$

where $u_j = u_j(\mathbf{x}, t)$ are the system variables (also called "species" or fields), $\mathbf{x} = (x_1, x_2)$ denotes the spatial coordinates and $\Delta u_j = \partial^2 u_j / \partial x_1^2 + \partial^2 u_j / \partial x_2^2$ are the second spatial derivatives in the diffusion terms providing spatial coupling and establishing (11.1) as a PDE. The values of the diffusion constants $d_1 \geq 0$ and $d_2 \geq 0$ depend on the spatial scale. Rescaling $x_j \mapsto cx_j$ results in $d_j \Delta u_j \mapsto d_j c^2 \Delta u_j$ and we can choose $c^2 = 1/d_1$ to set the diffusion coefficient of the first variable to 1 and obtain for the second diffusion coefficient $d = d_2/d_1$. Some exemplary time evolutions of the spatially extended Brusselator are shown in Figs. 11.1 and 11.3.

In the following we will continue the discussion with a general reaction-diffusion system with two fields $u_1$ and $u_2$ defined in a spatial domain $\mathcal{B}$, which can be written as:

$$\dot{u}_1 = f_1(u_1, u_2) + \Delta u_1$$
$$\dot{u}_2 = f_2(u_1, u_2) + d \Delta u_2 \tag{11.2}$$

or

$$\dot{\mathbf{u}} = \mathbf{f}(\mathbf{u}) + C \Delta \mathbf{u} \qquad (11.3)$$

with $\mathbf{u} = (u_1, u_2)$, $\mathbf{f} = (f_1, f_2)$, $C = \begin{pmatrix} 1 & 0 \\ 0 & d \end{pmatrix}$ and $\Delta \mathbf{u} = (\Delta u_1, \Delta u_2)$.

We are interested to know under what conditions such systems will develop pattern formation. Let $\mathbf{u}^* = (u_1^*, u_2^*)$ be the *uniform steady state* of the system (11.3). It satisfies $\nabla \mathbf{u}^* = 0 = \partial \mathbf{u}/\partial t$ and so also $f_1(u_1^*, u_2^*) = 0 = f_2(u_1^*, u_2^*)$. This state is the equivalent of fixed points for non-spatial systems (Sect. 1.4). If the uniform steady state (which has no patterns) is the only stable solution, i.e., the only attractor of the system, then there will be no pattern formation.[1] In this case, a mandatory condition for pattern formation is that the uniform steady state is unstable. This can be analyzed via linear stability analysis.

## 11.1.2 Linear Stability Analysis of Spatiotemporal Systems

To investigate the stability of the uniform steady state $\mathbf{u}^*$ we consider an infinitesimally small perturbation $\mathbf{v}$ of $\mathbf{u}^*$, $\mathbf{u}(\mathbf{x}, t) = \mathbf{u}^* + \mathbf{v}(\mathbf{x}, t)$ whose time evolution is given by the linear PDE

$$\dot{\mathbf{v}} = A\mathbf{v} + C \Delta \mathbf{v} \qquad (11.4)$$

where $A$ denotes the Jacobian matrix $J_{\mathbf{f}}(\mathbf{u}^*)$ of $\mathbf{f}$ at the steady state $\mathbf{u}^*$. Similarly to Sect. 1.4, we did a Taylor expansion keeping only linear terms. Due to linearity, any solution $\mathbf{v}(\mathbf{x}, t)$ of this linear PDE can be decomposed into Fourier modes $\mathbf{v}(\mathbf{x}, t) = \mathbf{v}_0 e^{\sigma t} e^{\langle \boldsymbol{\gamma}, \mathbf{x} \rangle}$ where $\langle \boldsymbol{\gamma}, \mathbf{x} \rangle = \gamma_1 x_1 + \gamma_2 x_2$ stands for the standard dot product. Substitution of $\mathbf{v}(\mathbf{x}, t)$ in (11.4) yields the algebraic equation

$$\sigma \mathbf{v} = B(\boldsymbol{\gamma})\mathbf{v} \qquad (11.5)$$

with

$$B(\boldsymbol{\gamma}) = A + \begin{pmatrix} \gamma_1^2 + \gamma_2^2 & 0 \\ 0 & d(\gamma_1^2 + \gamma_2^2) \end{pmatrix}.$$

Equation (11.5) defines an eigenvalue problem for the $(2, 2)$ matrix $B(\boldsymbol{\gamma})$ with eigenvalues $\sigma_{1,2}(\boldsymbol{\gamma}) = \lambda_{1,2} + i\omega_{1,2} \in \mathbb{C}$ and corresponding eigenvectors $\mathbf{e}_{1,2}(\boldsymbol{\gamma}) \in \mathbb{C}^2$. The homogeneous steady state $\mathbf{u}^*$ is stable if both real parts $\lambda_1$ and $\lambda_2$ are negative and thus the amplitude of the perturbation decays. If the imaginary parts $\omega_{1,2} = \pm\omega$ are not vanishing the perturbation increases or decreases in an oscillatory manner with frequency $\omega$.

---

[1] Of course, in general besides the homogeneous state there could be additional coexisting attractors in some extended systems.

The matrix $B$ (and thus also its eigenvalues and eigenvectors) depends on $\boldsymbol{\gamma} = (\gamma_1, \gamma_2) \in \mathbb{C}^2$ but not all values of $\boldsymbol{\gamma}$ are compatible with the respective *boundary conditions*.[2] To give examples for this selection we consider the dynamics in a two-dimensional domain $\mathcal{B} = \{\mathbf{x} = (x_1, x_2) \in \mathbb{R}^2 : 0 \leq x_1 \leq L_1 \wedge 0 \leq x_2 \leq L_2\}$.

Periodic boundary conditions $\mathbf{v}(x_1, x_2, t) = \mathbf{v}(x_1 + L_1, x_2, t)$ along the $x_1$ axis, for example, imply $e^{\gamma_1 L_1} = 1$ and thus $\gamma_1 = ik_1$ with $k_1 = m_1 2\pi/L_1$ and $m_1 \in \mathbb{Z}$. $k_1$ is the *wavenumber* in $x_1$-direction and $\ell_1 = 2\pi/k_1 = L_1/m_1$ is the corresponding wavelength. With periodic boundary conditions in both directions the matrix $B(\boldsymbol{\gamma})$ thus reads

$$B(k_1, k_2) = A - \begin{pmatrix} k_1^2 + k_2^2 & 0 \\ 0 & d(k_1^2 + k_2^2) \end{pmatrix}. \tag{11.6}$$

Another important case are *no-flux* or *absorbing boundary conditions*[3] $\mathbf{n} \cdot \nabla \mathbf{u} = 0$ where $\mathbf{n}$ denotes the outer normal to the boundary $\partial \mathcal{B}$ of the considered domain $\mathcal{B}$. For rectangular $\mathcal{B}$ these boundary conditions are given by

$$\frac{\partial u_j}{\partial x_1}(0, x_2, t) = \frac{\partial u_j}{\partial x_1}(L_1, x_2, t) = 0 \text{ and } \frac{\partial u_j}{\partial x_2}(x_1, 0, t) = \frac{\partial u_j}{\partial x_2}(x_1, L_2, t) = 0.$$

Since $\nabla \mathbf{u}^* = 0$ for the uniform state, the no-flux boundary conditions also hold for the perturbation $\mathbf{v}$ and the only Fourier modes fulfilling them are

$$\mathbf{v}(\mathbf{x}, y) = \mathbf{v}_0 e^{\sigma t} \cos(k_1 x_1) \cos(k_2 x_2) \tag{11.7}$$

with $k_j = m_j \pi/L_j$. Substituting this solution in (11.4) yields the same matrix $B$ as shown in (11.6), the only difference is the set of admissible wavenumbers which is $k_j = m_j 2\pi/L_j$ for periodic boundary conditions but $k_j = m_j \pi/L_j$ in case of no-flux conditions (with $m_j \in \mathbb{Z}$ in both cases).

The linear stability analysis of the steady state therefore in both cases consists of identifying ranges of values $q$ where the matrix

$$B(q) = A - \begin{pmatrix} q^2 & 0 \\ 0 & dq^2 \end{pmatrix} \tag{11.8}$$

has eigenvalues with negative or positive real parts. Note that this analysis is the same for 1D ($q^2 = k^2$), 2D ($q^2 = k_1^2 + k_2^2$), or 3D ($q^2 = k_1^2 + k_2^2 + k_3^2$) spatial domains.

All eigenvalues $\mu_{1,2} = \tau/2 \pm \sqrt{(\tau/2)^2 - \delta}$ of $B(q)$ have negative real parts if

$$\tau(q) = \text{trace}(A) - (1 + d)q^2 < 0 \tag{11.9}$$

$$\delta(q) = \det(A) - (dA_{11} + A_{22})q^2 + dq^4 > 0. \tag{11.10}$$

---

[2] Boundary conditions are the values the variables $u_i$ take or are forced to take at the boundaries of the spatial domain and may have a major impact on the dynamics.

[3] This is a special case of Neumann boundary conditions, $\mathbf{n} \cdot \nabla \mathbf{u} = \text{const.}$.

If the underlying system without diffusion is stable, then $\mathrm{trace}(A) < 0$ and therefore condition (11.9) is fulfilled for all $q$, because $d > 0$.

### 11.1.3 Pattern Formation in the Brusselator

To give concrete examples of pattern formation we return to the Brusselator in a square region of size $L_1 \times L_2 = 50 \times 50$

$$\begin{aligned}
\dot{u}_1 &= 1 - (b+1)u_1 + au_1^2 u_2 + \Delta u_1 \\
\dot{u}_2 &= bu_1 - au_1^2 u_2 + d\Delta u_2.
\end{aligned} \tag{11.11}$$

The Jacobian matrix at the steady state $\mathbf{u}^* = (1, b/a)$ is given by

$$A = J_f(\mathbf{u}^*) = \begin{pmatrix} b-1 & a \\ -b & -a \end{pmatrix} \tag{11.12}$$

and the maxima of the real parts $\mathrm{Re}(\lambda)$ of the eigenvalues of the matrix $B(q)$ (11.8) as a function of $q$ are shown in Fig. 11.2a.

The parameters $a = 9$ and $b = 9.8$ are chosen such that for $d = 1$ (i.e., for uniform diffusion) the system is linearly stable. If the diffusion constant $d$ of $u_2$ increases and exceeds some threshold a range of unstable $q$-values occurs where $\max(\mathrm{Re}(\lambda))$ is positive enabling growth of perturbations and pattern formation. This is the *Turing instability* which occurs if the inhibitor $u_2$ diffuses more rapidly than the activator $u_1$. Figure 11.1 shows pattern formation due to Turing instability for different values of the diffusion constant $d$. Starting from random initial conditions patterns quickly emerge. Some even arrange into perfectly symmetric patterns (like d, h).

So far we have considered the case where the steady state of the Brusselator is stable for small wave numbers $q$. Now we increase the parameter $b$ from $b = 9.8$ to $b = 10.2$ such that a long wavelength (i.e., small $q$) instability can occur as can be

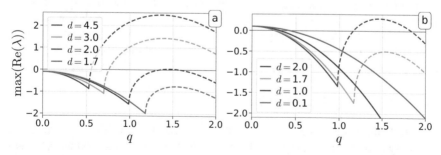

**Fig. 11.2** Maxima of real parts of eigenvalues of the matrix $B(q)$ versus $q$ for different diffusion constants $d$. Solid and dashed curves represent complex and real eigenvalues, respectively. **a** $a = 9$ and $b = 9.8$ and **b** $a = 9$ and $b = 10.2$

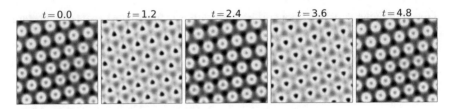

**Fig. 11.3** Snapshots of the $u_1$ variable of the Brusselator system (11.11) showing an oscillating pattern for $a = 9, b = 10.2$ and $d = 1.7$. A random initial condition was first evolved for $t = 1000$ units to reach the stable oscillating pattern. Animations of the time evolution can be found online at animations/11/brusselator

seen in Fig. 11.2b. Let's focus on the case of $d = 1.7$ where there is no actual Turing instability, because the dashed part of the curve is below 0. In this case, we would expect the emerging patterns to be uniform and oscillating in time from the point of linear stability analysis (uniform due to the largest Re($\mu$) being at $q = 0$, and oscillating because $\mu$ is complex). Nevertheless, the Turing mechanism still impacts the resulting patterns as can be seen in Fig. 11.3, which shows *oscillating* patterns instead of the static patterns of Fig. 11.1.

To summarize, in the linear stability analysis of the Brusselator we have seen two mechanisms leading to an instability of the uniform steady state: if condition (11.9) is violated for some wave number $q$ then a Hopf bifurcation takes place. This happens if $\max_q(\tau(q)) = \text{trace}(A) = b - 1 - a > 0$. The second instability mechanism is the Turing instability which occurs if condition (11.10) is not fulfilled. This is the case if $\min_q(\delta(q)) = \det(A) - (dA_{11} + A_{22})^2/4d = a - (db - d - a)^2/4d < 0$.

### 11.1.4  Numerical Solution of PDEs Using Finite Differences

The temporal evolutions shown in Figs. 11.1 and 11.3 were computed with the finite differences method, where spatial derivatives are approximated by difference quotients. In this way PDEs are converted to (large) sets of ODEs that can be solved by any ODE solver. Let $\mathbf{x}_{m_1, m_2} = (a_1 + (m_1 - 1)h_1, a_2 + (m_2 - 1)h_2)$ be a point on a grid of dimension $(M_1, M_2)$ covering the spatial domain $\{(x_1, x_2) : a_1 \leq x_1 \leq b_1 \wedge a_2 \leq x_2 \leq b_2\}$ with $h_i = (a_i - b_i)/(M_i - 1)$ being the spatial discretization length. The variables $\mathbf{u} = (u_1, u_2)$ evaluated at the grid point $(m_1, m_2)$ are given by

$$\mathbf{u}_{m_1, m_2}(t) = (u_1(\mathbf{x}_{m_1, m_2}, t), u_2(\mathbf{x}_{m_1, m_2}, t)). \qquad (11.13)$$

Of course, this discretization must also meet the boundary conditions. For periodic boundary conditions, for example, this can be achieved using the modulo operation $m_i = 1 + \text{mod}(m_i - 1, M_i - 1)$.

To simplify the formal description we use $M_1 = M_2 = M$ and consider a domain specified by $a_1 = a_2 = 0$ and $b_1 = b_2 = L$. In this case $h_1 = h_2 = h = L/(M - 1)$

**Fig. 11.4** Spiral wave in a 2D excitable medium. Regions in light blue are excitable and in dark blue/black refractory. The pink arrows indicate the direction of the motion of the wave front towards the excitable region. The region the wave had passed immediately before this snapshot is refractory and therefore does not support any wave propagation yet. As a result, a rotating wave occurs

and the lowest order approximation of the second derivatives in the diffusion term is given by the so-called *five point stencil*

$$\Delta \mathbf{u}_{m_1,m_2} = \frac{\mathbf{u}_{m_1+1,m_2} + \mathbf{u}_{m_1-1,m_2} + \mathbf{u}_{m_1,m_2+1} + \mathbf{u}_{m_1,m_2-1} - 4\mathbf{u}_{m_1,m_2}}{h^2}.$$

With this definition the ODEs for the variables at the $(M, M)$ grid points are

$$\dot{\mathbf{u}}_{m_1,m_2} = \mathbf{f}(\mathbf{u}_{m_1,m_2}) + \begin{pmatrix} 1 & 0 \\ 0 & d \end{pmatrix} \Delta \mathbf{u}_{m_1,m_2}.$$

## 11.2  Excitable Media and Spiral Waves

Excitable media are spatially extended excitable systems (see Sect. 2.2.4) that can be found almost everywhere in nature. An illustrative example is a wildfire. The forest can be locally excited (i.e., set on fire) resulting in flame fronts leaving behind burned vegetation. The vegetation can carry another flame front only after it has regrown again. The time trees and bushes need to recover is the *refractory time* or *period*. After a wave (here: the flame front) has passed some location the next wave can therefore run across the same spot only after the refractory period has passed. The existence of a refractory phase means that an excitation wave cannot propagate in any direction but only to the excitable region of the medium. As a result, rotating waves, also called *spiral waves* may occur as illustrated in Fig. 11.4. Spiral waves are omnipresent in excitable media and can be seen, for example, not only in chemical reactions like the Belousov-Zhabotinsky reaction, but also in cardiac and neural tissue or colonies of slime moulds. An example animation can be found in animations/11/fitzhugh_spiralwave.

## 11.2.1   The Spatiotemporal Fitzhugh-Nagumo Model

A fundamental model describing an excitable medium is the FitzHugh-Nagumo model (2.6), extended in space with a diffusive coupling in the fast variable $u$

$$\dot{u} = au(u - b)(1 - u) - w + d\Delta u$$
$$\dot{w} = \varepsilon(u - w).$$

(11.14)

The diffusion constant $d$ can be set to 1 by rescaling $x_j \mapsto x_j/\sqrt{d}$ the spatial coordinates $(x_1, x_2)$. Figure 11.5 shows differently initialized waves in a domain of size $L = 300$ with no-flux boundary conditions. In the first row two concentric waves emerge from two local point-like stimuli. In contrast to wave solutions of linear wave equations, colliding waves in excitable media cannot pass through each other but instead annihilate each other. Once the waves reach the boundaries they disappear due to the no-flux boundary conditions. Dark colors inside the circular waves indicate refractory zones where no further excitation of the medium is possible (at that time). Once the light blue colors reappear indicating the value of the resting state $u = 0$ the refractory period is over and the system is locally ready again for the next excitation.[4]

The second example consists of a first perturbation along a vertical line on the left at $t = 0$ (red line in Fig. 11.5e) which generates two plane waves running to the left and to the right, where the wave to the left is immediately absorbed by the no-flux boundary. Once the plane wave propagating to the right entered the right half of the domain a horizontal rectangular region (red bar in Fig. 11.5f) is excited in the wake of the wave at $t = 300$. Since the region between this excitation and the propagating plane wave is still refractory the second excitation can only spread into non-refractory regions. This leads to the formation of two counter rotating spiral waves as shown in the snapshots at $t = 400$ and $t = 780$ in Fig. 11.5g, h, respectively. The rotation of these spirals leads to a self-exciting periodic generation of circular waves moving towards the boundary.

## 11.2.2   Phase Singularities and Contour Lines

The most important part of a spiral wave is the region around its tip where it is most susceptible to perturbations. While there is no general definition where exactly the tip is located there are essentially two groups of approaches that specify the center of a spiral wave. One is based on the concept of phase singularities and the other on the intersection of contour lines describing wavefronts.

A stable rotating spiral wave at a fixed location leads to periodic oscillations at each point $\mathbf{x}$ of the domain. As discussed in Sect. 2.2.2, periodic oscillations can

---

[4] In general, for systems with many coupled fields, one has to consider all of them together to deduce the range of excitability. For the Fitzhugh-Nagumo model we may focus on the value of $u$, only.

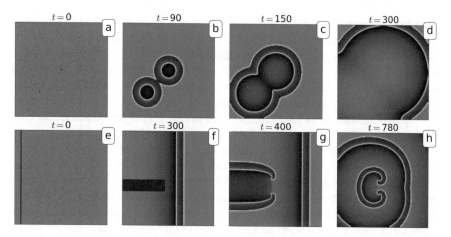

**Fig. 11.5** Snapshots of the $u$ variable of the 2D FitzHugh-Nagumo model (11.14) showing excitation waves for $a = 3$, $b = 0.2$, $\varepsilon = 0.01$ and $d = 1$. **a–d** Stimulation at two locations results in concentric waves. **e–h** A sequence of stimulations at vertical ($t = 0$) and horizontal ($t = 300$) stripes leads to counter-rotating spiral waves periodically ejecting concentric waves running to the boundary. An animation of the time evolution of the examples shown is provided online at animations/11/fitzhugh

always be uniquely described by a phase. Using both variables $u(\mathbf{x}, t)$ and $w(\mathbf{x}, t)$ a phase[5] can be computed as

$$\alpha(\mathbf{x}, t) = \arctan\left( \frac{w(\mathbf{x}, t) - \bar{w}}{\sigma_w}, \frac{u(\mathbf{x}, t) - \bar{u}}{\sigma_u} \right) \tag{11.15}$$

where $\bar{u}$ and $\bar{w}$ are mean values and $\sigma_u$ and $\sigma_v$ the corresponding standard deviations of $u$ and $w$, respectively.

In Fig. 11.6a, b snapshots of the corresponding fields $u(\mathbf{x}, t)$ and $w(\mathbf{x}, t)$ are plotted and Fig. 11.6c, d show color coded the resulting phase $\alpha(\mathbf{x}, t)$, also called *phase maps*, computed with (11.15). The phases at locations near the center of the spiral wave are displayed in the zoom in Fig. 11.6d. At the tip of the spiral a *phase singularity* occurs where the oscillation amplitudes of $u$ and $w$ vanish and the phase is thus not defined. Another feature of a phase singularity is that it is surrounded by all possible phase values $\alpha \in [-\pi, \pi)$. Therefore, a phase singularity results in a non-vanishing integral of the gradient of the phase along a closed counter clockwise path $\partial \mathcal{R}$ encircling a (small) region $\mathcal{R} \subset \mathbb{R}^2$ containing the phase singularity. If several phase singularities are located in this region the integral

$$c(t) = \frac{1}{2\pi} \oint_{\partial \mathcal{R}} \nabla \alpha(\mathbf{x}, t) \, d\mathbf{x} = n - m \tag{11.16}$$

---

[5] To be a bit more precise, $\alpha$ is a protophase according to Sect. 2.2.2 but for simplicity here we will call it phase.

gives the numbers $n$ and $m$ of counterclockwise and clockwise rotating phase singularities in $\mathcal{R}$, respectively.[6] In Fig. 11.6 we have $n = 0$ and $m = 1$. $c = n - m$ is also called the *topological charge* that can change only if phase singularities enter or leave $\mathcal{R}$. It does not change if a new pair of counter rotating spiral waves (with $n = 1 = m$) is created in $\mathcal{R}$ or if the tips of two such waves in $\mathcal{R}$ collide and annihilate each other. The exact location of a phase singularity depends on the definition of the phase variable used, but differences are practically negligible and in any case a point very close to the center of the spiral wave is obtained.

The second approach to specify the tip of the spiral is based on contour lines of some observables. For example, you can determine curves with $\dot{w} = 0$ which can be considered as the boundary of the excited region. They represent the wave front *and* the wave back which meet in the tip of the spiral (see Fig. 11.6b). The curve $\dot{u} = 0$ separates wave front and wave back, and both curves intersect in the center of the spiral wave as illustrated in Fig. 11.6d. Another potential choice of contour lines would be $u - \bar{u} = 0$ and $w - \bar{w} = 0$. In case only one variable $u$ is available due to, e.g., experimental measurement, you can create a second variable via delay embedding (Chap. 6), e.g., $u(\mathbf{x}, t - \tau_d)$. Then you can compute phase angles using $u(t), u(t - \tau_d)$ in 2D delay embedding space (at each point in the spatial domain). Alternatively, you can look for intersections of contour lines derived from the observed quantity $u$ (see references in Further reading).

Depending on initial conditions several coexisting spiral waves may occur at different locations and they may move around. Such meandering can be illustrated, e.g., by tracing the locations of phase singularities. Rotating waves in 3D excitable media are called *scroll waves* and their centers are characterized by phase singularities constituting filaments (i.e., line segments). An example of such 3D scroll waves can be seen online on animations/11/scroll_wave Note, however, that in any case the

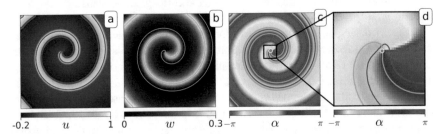

**Fig. 11.6** Snapshots of a clockwise rotating spiral wave generated with the 2D FitzHugh-Nagumo model (11.14) for $a = 3$, $b = 0.2$, $\varepsilon = 0.01$, $d = 1$ and a domain size of $300 \times 300$ covered by $150 \times 150$ grid points. **a** and **b** show color coded the variables $u$ and $w$, respectively. The resulting phases $\alpha(\mathbf{x}, t)$ are shown in **c**. **d** is a zoom of (**c**) at the phase singularity, a point surrounded by all possible phase values $\alpha \in [-\pi, \pi]$. The purple and cyan contour lines indicate locations with $\dot{u} = 0$ and $\dot{w} = 0$, respectively, and intersect close to the phase singularity marked by a white circle

---

[6] Stokes' theorem implies that $c(t)$ is zero if $\alpha(\mathbf{x}, t)$ is continuously differentiable in $\mathcal{R}$. This is not the case if $\mathcal{R}$ contains a phase singularity.

assumption of an approximately periodic motion has to be fulfilled to define local phases. This is typically not the case with highly chaotic wave dynamics that we shall see in Sect. 11.3.3.

## 11.3 Spatiotemporal Chaos

Spatially extended systems may not only exhibit static or periodic oscillation patterns or (spiral) waves but also chaotic dynamics. In that case wave-like structures appear, move around, merge or split, and disappear in a rather complicated and unpredictable manner. And in most systems the dynamics becomes more complex the larger the spatial extend of the system is, as we will see in the following subsections.

### 11.3.1 Extensive Chaos and Fractal Dimension

What is the state space dimensionality $D$ of a spatiotemporal system? Well, the formal answer here is that it is *infinite*, as the spatial coordinate $x$ has an infinite amount of values and states are represented by functions of space. Of course, when we simulate these systems, the space is discretized and then the state space dimension $D$ is large but finite because it is the same as the amount of grid points in the spatial discretization, multiplied by the number of coupled fields.

In general however, more important than the dimension of the state space is the actual fractal dimension of the attractor (Sect. 5.4), that can be specified by the Lyapunov dimension $\Delta^{(L)}$, (5.9). In most spatially extended chaotic systems $\Delta^{(L)}$ increases with the system size. This feature is called *extensive chaos*. Intuitively you may understand this phenomenon by considering the full system as a combination of weakly coupled compartments that evolve essentially along their "own" chaotic evolution. Within this picture, making the system larger means to add more compartments or chaotic units and in this way the number of positive Lyapunov exponents increases. On the other side, what makes the fractal dimension a meaningful quantifier is that its value typically converges as we make the spatial discretization more and more fine (which would also increase the state space dimension $D$, just like increasing the system size would).

To demonstrate these effects we consider the *Kuramoto-Sivashinsky equation*[7]

$$\frac{\partial u}{\partial t} = -\frac{\partial^2 u}{\partial x^2} - \frac{\partial^4 u}{\partial x^4} - u\frac{\partial u}{\partial x} \tag{11.17}$$

which is described by a single variable $u(x, t)$ in a one dimensional domain $\mathcal{B} = \{x \in \mathbb{R} : 0 \leq x \leq b\}$. Originally this PDE was introduced as a model for flame fronts but

---

[7] In the literature you may also find the notation $u_t = -u_{xx} - u_{xxxx} - uu_x$.

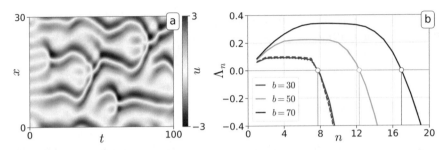

**Fig. 11.7** **a** Spatiotemporal chaos in the Kuramoto-Sivashinsky equation (11.17) with periodic boundaries and domain size $b = 30$. **b** Cumulative sum of Lyapunov exponents of the system for different domain sizes $b$. Solid lines are with $h = 0.3$, and for purple ($b = 30$) we have additionally dashed $h = 0.25$ and dotted $h = 0.35$. Vertical lines give the Lyapunov dimension (Sect. 5.4.4) for each, $\Delta^{(L)} = 31.76, 57.19$, and $85.86$

in the past decades it became a famous test case for illustrating novel concepts dealing with highdimensional and rather complex dynamics. The boundary conditions for this system are typically taken as periodic.

The uniform steady state $u^* = $ const. is a solution of this PDE. Once the system size is increased for $b \geq 2\pi$ the steady state $u^*$ becomes unstable and static patterns as well as travelling waves occur. For even larger domains further bifurcations and chaotic dynamics appear. Figure 11.7a shows a *space-time plot* of $u$, showing spatiotemporal chaos.

In Fig. 11.7b we present a plot of the cumulative sum of the Lyapunov exponents of the Kuramoto-Sivashinsky system, $\Lambda_n = \sum_{i=1}^{n} \lambda_i$ versus $n$. The value of $n$ for that $\Lambda_n$ crosses 0 is the Kaplan-Yorke dimension $\Delta^{(L)}$. It becomes apparent both that $\Delta^{(L)}$ increases with $b$, but also that it converges when the spatial discretization $h$ becomes finer.

### 11.3.2  Numerical Solution of PDEs in Spectral Space

The method we outlined in Sect. 11.1.4, in combination with the algorithm of Sect. 3.2.2, can be used to calculate the Lyapunov spectrum of the Kuramoto-Sivashinsky system. Well, it can in theory, but in practice it would take almost forever to do so, because to avoid numerical instabilities you would have to use a very fine spatial discretization with many grid points. For some PDEs a faster alternative is to solve them in *spectral space* (i.e., the Fourier space), which is efficient for PDEs that have a small amount of nonlinearity but higher order spatial derivatives.

Let $y(k, t) = \mathcal{F}(u) = \int e^{-ikx} u(x, t)\, dx$ be the Fourier transform of $u$. For any linear term, e.g., $\partial^2 u / \partial x^2$, we have straightforwardly that $\mathcal{F}(\partial^2 u / \partial x^2) = \partial^2 (\mathcal{F}(u)) / \partial x^2 = (-ik)^2 \mathcal{F}(u) = -k^2 y$, as linear operations can be interchanged and both $\partial^2 (\cdot)/\partial x^2$ and $\mathcal{F}(\cdot)$ are linear. So, the spatial derivative in terms of $u$ has been changed into a trivial multiplication in terms of $y$! The tricky part of course are nonlinear terms like $u\partial u/\partial x$, and these can only be treated on a case-by-case approach. Here, we can re-write $u\partial u/\partial x = (1/2)\partial(u^2)/\partial x$, and then we have that $\mathcal{F}(u\partial u/\partial x) = (-ik/2)\mathcal{F}(u^2)$. In the end, (11.17) becomes simply

$$\frac{dy}{dt} = \frac{ik}{2}\mathcal{F}(u^2) + (k^2 - k^4)y. \tag{11.18}$$

To compute $\mathcal{F}(u^2)$ one needs to first calculate $u = \mathcal{F}^{-1}(y)$, square it, and then transform back, at every step of the evaluation of $dy/dt$. Solving (11.18) numerically is much faster than solving (11.17) with a spatial discretization. An optimized code example is shown in Code 11.1.

**Code 11.1** High-performance solving of the Kuramoto-Sivashinsky system in spectral space (Eq. (11.18)) using planned Fourier transforms.

```
using OrdinaryDiffEq, LinearAlgebra, FFTW
b, dx = 30, 0.2 # domain size, spatial discretization
x = range(0, b; step = dx) # real space
u0 = @. cos((2π/b)*x) # initial condition
ks = Vector(FFTW.rfftfreq(length(u0))/dx) *2π # wavenumbers
# Pre-caculate Fourier transforms and other factors
forw_plan = FFTW.plan_rfft(u0)
inv_plan  = FFTW.plan_irfft(y0, length(u0))
y0 = forw_plan * u0
ik2 = @. im*ks/2; k²_k⁴ = @. ks^2 - ks^4
ydummy = copy(y0); udummy = copy(u0)
p = (forw_plan, inv_plan, udummy, ydummy, k²_k⁴, ik2)

function kse_spectral!(dy, y, p, t)
    forw_plan, inv_plan, udummy, ydummy, k²_k⁴, ik2 = p
    y² = begin # nonlinear term
        mul!(udummy, inv_plan, y)       # transform to u
        udummy .*= udummy               # square u
        mul!(ydummy, forw_plan, udummy) # transform to y
    end
    @. dy = y²*ik2 + k²_k⁴*y # KS equation, spectral space
end
prob = ODEProblem(kse_spectral!, y0, (0.0, 1000.0), p)
sol = solve(prob, Tsit5(); saveat = 0:0.5:1000.0)
u = [inv_plan*y for y in sol.u] # solution in real space
```

## 11.3.3   Chaotic Spiral Waves and Cardiac Arrhythmias

Spatiotemporal chaos can also occur in reaction-diffusion systems and many other extended systems. In the following we will focus on excitable media with unstable spiral waves and we will discuss their chaotic dynamics in the context of cardiac arrhythmias.

The heart muscle consists of excitable heart muscle cells (cardiomyocytes) that represent (when seen as a continuum) an excitable medium. The normal heart beat, triggered by the sinu-atrial node acting as a pacemaker, essentially consists of plane waves of the transmembrane voltage resulting in coordinated contractions of the heart. This pumping functionality is impaired if spiral waves occur, detected in elec-trocardiograms (ECGs) as periodic signals with a relatively high frequency indicating a so-called *tachycardia*. If, due to any perturbation or changes in electrophysiology the excitation waves turn into a chaotic dynamics with multiple wave collisions and break-ups, the heart is in a state of atrial or ventricular fibrillation, where it fails to pump blood properly. Ventricular fibrillation is lethal and has to be terminated imme-diately. Currently, the only method to do this is defibrillation, where strong electrical shocks are applied, pushing the medium (i.e., all cardiomyocytes) to its refectory state. Once the refractory period is over the sinu-atrial node can take control again and restore a normal heart beat.

Full modelling of cardiac electrophysiology requires detailed models with many ODEs for the transmembrane voltage and different transmembrane ionic currents controlled by gating variables. A qualitative dynamical understanding, however, can already be achieved with simplified models consisting of differential equations for a fast variable representing the cell membrane voltage and for at least one additional variable standing for the ionic currents. An example for this qualitative approach is the *Aliev-Panfilov model*

$$\dot{u} = ku(1 - u)(u - a) - uw + d\,\Delta u$$
$$\dot{w} = \left(\varepsilon + \frac{\mu_1 w}{\mu_2 + u}\right)[-w - ku(u - a - 1)] \tag{11.19}$$

that describes an excitable cardiac medium. Depending on parameter values and initialization, stable concentric, planar, or spiral waves can occur, like those we have seen with the FitzHugh-Nagumo model (Fig. 11.5). In addition, this system may also exhibit spatiotemporal chaos. For example, with parameter values $k = 8$, $a = 0.05$, $\varepsilon = 0.002$, $\mu_1 = 0.2$, $\mu_2 = 0.3$, $d = 0.2$, a domain size of $L = 100$ and (400,400) grid points the initialization of a single spiral wave results via wave breakup(s) in the complex wave patterns shown in Fig. 11.8. In the course of the chaotic evolution spiral waves (or wavelets) rotating in opposite directions are continuously created or disappear due to pairwise annihilation or when running into the no-flux boundary. For this system the local dynamics at any grid point **x** is so chaotic that it cannot be (approximately) described by a phase, anymore.

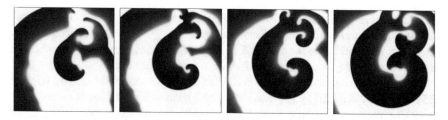

**Fig. 11.8** Consecutive snapshots of the chaotic evolution of the wave dynamics $u(\mathbf{x}, t)$ of the Aliev-Panfilov model (11.19) sampled with $\Delta t = 10$

Actually, the spatiotemporal wave chaos shown in Fig. 11.8 doesn't last forever. If you wait long enough, at some moment in time—that is very, very difficult to predict—a configuration occurs with a single wave disappearing in the boundary. What remains is a quiescent medium. Therefore (for the parameters chosen) the Aliev-Panfilov model is not exhibiting persistent spatiotemporal chaos like the Kuramoto-Shivashinsky equation of the previous subsection. Rather, it displays a form of *transient chaos* which occurs with many spatially extended systems. In general the average period of time where the system is chaotic increases exponentially with the system size. If the mean chaotic transient lasts longer than any (feasible or affordable) simulation the question whether the system is permanently chaotic with a chaotic attractor, or just exhibits a chaotic transient, can not be answered by just simulating the dynamical equations. Furthermore, in some cases the mean chaotic transients can be so long that distinguishing between persistent chaos and transient chaos may not even have any practical impact.

## Further Reading

The review article on pattern formation by Cross and Hohenberg [242] from 1993 still provides a good overview with many details and examples. Worth reading are also the books by Cross and Greenside [243] and Murray [244] which cover a broad range of topics on dynamics of extended systems including excitable media. For more background on the Brusselator see also the Further reading section of Chap. 2.

Turing patterns are named after Alan Turing who described in 1952 in his seminal paper "The Chemical Basis of Morphogenesis" [245] the instability mechanism due to different diffusion constants. The Turing instability was studied by many authors. Work that inspired the presentation in this book was published by Yang and Epstein [246].

Spiral waves in excitable media have been observed in chemical systems like the Belousov-Zhabotinsky reaction [247, 248], slime mold colonies [249] and catalytic CO oxidation on platinum surfaces [250], to mention only a few. Their importance for cardiac dynamics has first been pointed out by Krinsky [251] and Winfree [252].

An overview and classification of models describing cardiac electrophysiology is given in the Scholarpedia article "Models of cardiac cell" [253] curated by Fenton and Cherry. The same authors also run a webpage called "The Virtual Heart" http://thevirtualheart.org/ with many animations including WebGL applications for fast interactive simulations on the user's GPU. Implementations of many cell models can also be found in the repository CellML https://www.cellml.org/

Phase singularities and their role in cardiac dynamics were discussed by Winfree [254] and Gray et al. [255]. Iyer and Gray [256] address localization accuracy when using line integrals and phase maps. Spiral tip detection using contour lines and iso-surfaces has be used and discussed by Barkley et al. [257], Holden and Zhang [258], Fenton and Karma [259], Marcotte and Grigoriev [260] and Gurevich and Grigoriev [261].

Excitation waves in 3D are called *scroll waves* and they are characterized by onedimensional phase filaments. These filaments play a major role in understanding cardiac dynamics [262, 263] and their motion is affected by heterogeneities in the medium as discussed, e.g., in [264] (and further references therein). Since the heart is not only excitable but also contracting mechanical spiral waves have also been observed in the heart [265].

In some simulations chaotic spiral wave dynamics in excitable media has been found to be transient chaos [266], an observation that is confirmed by clinical observations of self-termination of arrhythmias (see [267–269] and references there in). If no self-termination occurs, in case of ventricular fibrillation immediate treatment using strong electrical defibrillation shocks is mandatory, but there is also ongoing research on low-energy defibrilation methods [270].

The Kuramoto-Sivashinsky equation is named after Y. Kuramoto and G. Sivashinsky who studied (independently) turbulent distributed chemical reactions [271] and instabilities in flame fronts [272, 273], see also [274].

## Selected Exercises

11.1 How many uniform steady states possesses the system $\partial_t u = ru - u^3 + \partial_x^2 u$? For each uniform steady state derive linearized evolution equations for infinitesimal perturbations $v(x, t)$ and perform a linear stability analysis for periodic boundary conditions. Find the critical values $r_c$ of the parameter $r$ for which each uniform steady state becomes unstable and the wavenumber of the corresponding fastest growing mode. Plot the growth rate $\sigma(q)$ versus $q$.

11.2 Perform a linear stability analysis of the uniform steady state of the Swift-Hohenberg model $\partial_t u = (r - 1)u - 2\partial_x^2 u - \partial_x^4 u - u^3$ in a 1D domain $-L/2 \leq x \leq L/2$ with periodic boundary conditions. Determine the growth rate $\sigma$ of linear modes as a function of the control parameter $r$ and the wavenumber $q$. For which value $r_c$ do you find an instability? What is the wave number $q_c$ of the most unstable mode at the onset of the instability?

11.3 Use the finite differences method to solve the FitzHugh-Nagumo model (11.14) and reproduce Fig. 11.6. *Hint: use as initial condition: $u(x_1, x_2, 0) =$*

0.5 *for* $x_1 < 2/3$, $u(x_1, x_2, 0) = 0$ *for* $x_1 > 2/3$, $w(x_1, x_2, 0) = 0.5$ *for* $x_2 < 7/12$, $w(x_1, x_2, 0) = 0$ *for* $x_2 > 7/12$.

11.4 Using the same code simultaneously generate two plane waves that head towards each other (set $u = 0.5$ for a grid "line" that represents the start of the wave). Describe what happens once the waves reach each other.

11.5 Locate the phase singularity of a rotating spiral wave by numerically computing the topological charge (11.16) with shortest closed paths $(i, j) - (i, j + 1) - (i + 1, j + 1) - (i + 1, j) - (i, j)$ on the grid used to solve the PDE. Approximate the gradient by differences of phase values. *Hint: use* $\mathrm{diff}(\alpha_1 - \alpha_2) = \mathrm{mod}(\alpha_1 - \alpha_2 + \pi, 2\pi) - \pi$ *to compute phase differences taking into account phase jumps of $2\pi$ of the* arctan *function. Compare the location of the phase singularity with the intersection point of contour lines* $u - \bar{u} = 0$ *and* $w - \bar{w} = 0$, *where $\bar{u}$ and $\bar{w}$ are mean values of $u$ and $v$ respectively. Explain your result.*

11.6 Compute the time evolution of a spiral wave and use it to plot contour lines $u(\mathbf{x}, t) - u(\mathbf{x}, t + \tau_d) = 0$ and $u(\mathbf{x}, t) - u(\mathbf{x}, t - \tau_d) = 0$ for some suitable delay time $\tau_d$. Compare the intersection point with the location of the phase singularity at time $t$.

11.7 Show that the Kuramoto-Sivashinsky equation (11.17) with $0 \le x \le b$ is equivalent to the PDE $\frac{\partial u}{\partial t} = -r\frac{\partial^2 u}{\partial x^2} - \frac{\partial^4 u}{\partial x^4} - u\frac{\partial u}{\partial x}$ with $x \in [0, 1]$ and determine analytically the critical value of the parameter $r$ for which the steady state solution of this system becomes linearly unstable (for no-flux boundary conditions). *Hint: perform a linear stability analysis.*

11.8 Write code that solves the Kuramoto-Sivashinsky equation in spectral space, (11.18), and use it to produce space-time plots like Fig. 11.7a. Try to determine the critical domain size $b_c$ where asymptotic patterns start to oscillate.

11.9 Derive the equations for the linearized dynamics of the Kuramoto-Sivashinsky equation in spectral space. The approach is similar to that in Chap. 1 for deriving linearized dynamics, and then the resulting equations have to be transformed in spectral space. The final equation should read

$$\frac{d\delta y}{dt} = (k^2 - k^4)\delta y + ik\mathcal{F}\left(\mathcal{F}^{-1}(y)\mathcal{F}^{-1}(\delta y)\right)$$

with $y = \mathcal{F}(u)$ and $\delta y = \mathcal{F}(\delta u)$ the deviation vector(s) in spectral space.

11.10 Write code that solves the Kuramoto-Sivashinsky equation and its linearized dynamics in parallel in spectral space. Use the code to compute the 10% largest Lyapunov exponents using the algorithm of Chap. 3. Repeat the calculation as a function of the system size $b$. Which is the smallest $b$ where $\lambda_1 > 0$?

11.11 Simulate the Kuramoto-Sivashinsky equation with periodic boundary conditions in the chaotic regime, $b \ge 20$. Then, compute the Pearson correlation coefficient and mutual information between two spatial point timeseries as a function of their distance. How would you expect these coefficients to behave for increasing distance? Does your numerics confirm your conjecture? *Hint:*

*for better precision, you can average Pearson/MI over several pairs with same grid distance.*

11.12 Another prototypical example of an excitable medium is the *Barkley model* $\partial_t u = u(1 - u)(u - (w + b)/a) + d\Delta u$, $\partial_t w = \epsilon(u^\gamma - w)$ in 2D with non-flux boundary conditions. Determine the nullclines of the non-extended system ($d = 0$). Then simulated the model on a domain of size $300 \times 300$ with $(a, b, d, \epsilon) = (0.6, 0.05, 1., 0.02)$ for $\gamma = 1$ and $\gamma = 3$. Do you observe stable spiral waves or chaotic dynamics? Now change parameters to $(a, b, d, \epsilon) = (0.7, 0.05, 1, 0.08)$ and repeat the simulations. What do you see now for both ? values? *Hint: initialize both fields $u$ and $w$ with orthogonal stripes with zeros and ones.*

11.13 Produce an animation of Fig. 11.8 for both fields of the Aliev-Panfilov model. Repeat the process with increasing or decreasing the diffusion constant $d$.

11.14 Continue from the above exercise. At a given spatial pixel, plot the timeseries $u(t)$, $w(t)$ versus $t$ there. Also plot them as in a state space plot, $u(t)$ versus $w(t)$. Can you introduce some kind of approximate phase, like in the case of chaotic synchronization as in Sect. 9.3.1?

11.15 Calculate and plot curves with $du/dt = 0$ and $dw/dt = 0$ (i.e., nullclines) for the two fields of the Aliev-Panfilov over the field plots (like in Fig. 11.6). Can you calculate a phase map based on the phase $\alpha$ defined in (11.15)? Before applying (11.15) plot at some locations $\mathbf{x}_k$ in the 2D domain the projected trajectory $u(\mathbf{x}_k, t)$ versus $w(\mathbf{x}_k, t)$ for a period of time that is long compared to the typical time scales of this system. Can you determine a center that is suitable for computing $\alpha$?

# Chapter 12
# Nonlinear Dynamics in Weather and Climate

**Abstract** In this very last chapter we discuss concepts of nonlinear dynamics that are of high relevance for weather and climate. The first topic concerns the impact of sensitive dependence on initial conditions to the prediction of very complex systems like the weather. It turns out that the impact of deterministic chaos on the prediction of such systems is much milder than one would expect. Next on the list is a discussion of tipping points in dynamical systems, which describe when and how a system may tip from one stable state to another one. This topic is of crucial importance when one considers climate change scenarios. We close the chapter with an interesting presentation of various nonlinear dynamics applications in climate research, using examples paralleling what we've learned in this book so far.

## 12.1 Complex Systems, Chaos and Prediction

As of the publication date of this book, we can predict the weather with good accuracy for about 1–2 weeks. Given what you've learned so far, especially in Sect. 3.1.2 about the predictability horizon, this sounds really, really impossible. Let's do a back of the envelope estimate: we can safely assume that the most chaotic part of weather dynamics is atmospheric turbulence near the surface of the Earth, and estimates for its Lyapunov time $1/\lambda_1$ are about $1\,\mathrm{s}$. According to Sect. 3.1.2 the initial error $\varepsilon$ of determining the state of the entire weather system (which in principle has billions of degrees of freedom) will reach 1 in $\ln(1/\varepsilon)/\lambda_1$ seconds. Then naively we can say that our initial state error should be $\varepsilon \approx e^{-604800}$ to allow a satisfactory prediction of up to one week (1 week $= 604800\,\mathrm{s}$). It is beyond unreasonable to expect we can measure the entire weather system with such accuracy, so what's going on here?

Weather is a *complex* system, a term used to describe nonlinear dynamical systems that have multiple temporal and spatial scales and/or have a huge array of disparate interacting components. Weather, climate, societal systems, economic systems, power grids, real ecosystems, are all complex systems. All of these have either an infinite dimensional state space or at least an extremely high-dimensional one. Even so there are in fact several reasons for why surprisingly long time predictions are possible, even in the presence of deterministic chaos. Keep in mind that although the following discussion is based on weather, most of the concepts discussed can be applied to most complex systems.

**Errors beyond the linear regime.** The first, and most obvious, is that initial state errors $\varepsilon$ are well beyond the linearized regime the maximum Lyapunov exponent quantifies. Have a look at Fig. 3.2 for example, and consider the scenario that the initial error is already of order 1. Then, the error is already in the *saturation regime*, where the linearized dynamics does not hold anymore, and instead the nonlinear dynamics folds the state space back in. The error evolution is nonlinear and complicated, but its magnitude no longer increases exponentially.

**Sparse state space.** As we discussed in Chap. 11 spatiotemporal systems (PDEs) have in principle infinite-dimensional state spaces, but the effective dimensionality of motion quantified by the Lyapunov dimension, Sect. 5.4, often is finite and saturates with increasing spatial resolution. This has two consequences: (i) the effective dimension is much smaller than the billions of degrees of freedom a weather model may have. (ii) only a small portion of directions in the state space is unstable (expanding in the terms of Lyapunov exponents, Chap. 3), while the remaining are negative. As a result, a random perturbation will only have a tiny component in the direction(s) of maximal growth. Some time is therefore needed to align it to the maximally expanding directions. Thus, at least in the initial evolution of the system, perturbations do not yet follow the exponential growth characterized by the maximum Lyapunov exponent.

The following simple discrete system, which we will call "logistic transport", demonstrates well the ideas we just discussed

$$x_{n+1}^{(1)} = 4x_n^{(1)}(1 - x_n^{(1)}), \quad x_{n+1}^{(k)} = a_k x_n^{(k-1)}, \quad \text{for } k = 2, \dots, D \tag{12.1}$$

with $a_k$ chosen randomly and uniformly in the interval [0.87, 1.07] so that there are more contracting than expanding directions. The logistic transport has a tunable state space dimension $D$. It is chaotically driven by the first dimension $x^{(1)}$, which follows the dynamics of the logistic map for $r = 4$, (1.3). The remaining $D - 1$ maps are linear, the majority of which provide dissipation to the system, and their purpose is to emulate diffusion present in spatiotemporal (PDEs) systems. For any $D$, the maximum Lyapunov exponent of the system is $\log(2)$, equivalent to the Lyapunov exponent of the logistic map for $r = 4$.

In Fig. 12.1a we show how the fractal dimension of the underlying time evolution remains extremely small as we increase $D$. In Fig. 12.1b we show the instantaneous exponential growth rates of random perturbations. The curves are the result of averaging several thousand random perturbations over several thousand possible initial conditions. All perturbations will asymptotically grow exponentially with rate $\lambda_1 = \log(2)$. However, as it becomes clear from the figure, the actual growth rate starts much smaller than the expected maximum. In fact, when increasing $D$ really high, the initial perturbation growth rate is *negative* instead of positive, because the perturbation is aligned mostly with directions of shrinking in the state space, as these directions occupy the majority of possible directions. After some time evolution the perturbations align with the direction of maximal expansion. The larger the $D$, the more time this alignment requires. This effectively increases the prediction horizon of Sect. 3.1.2 without actually decreasing $\lambda_1$ nor increasing our measurement accuracy.

**Dynamics on multiple scales.** While it is clear that the most chaotic timescale of weather is seconds, weather is determined more by large scale structures, e.g., low and high pressure systems. The instability timescales of these structures are much larger than a second yet their impact on the evolution of the system is in a sense also much larger than that of the most chaotic scales. The conclusion is that, in general, the strongly chaotic motion of many small scale degrees of freedom has a much weaker effect on the large scale dynamics, than what would be naively expected from the Lyapunov time (i.e., that the weather would change dramatically within a couple of minutes due to exponential growth).

**Modelling errors act like stochastic noise.** Everything we've said so far is mostly theoretical reasoning because of the simple reason that it is computationally impossible to resolve the entire earth system on the scale of a meter and timescale of a millisecond. Therefore, the highly chaotic turbulent processes are not even resolved in weather models. The missing processes are therefore parameterized and only their average effect, instead of their fast fluctuations, is felt on the remaining scales that can be computationally resolved. The absence of the smallest scales can be approximated as noise input (also called stochastic driving) for the simulated system. This driving of course introduces errors that increase over time, but this increase is not exponential, but rather some low order polynomial function of time (e.g., linear or quadratic).

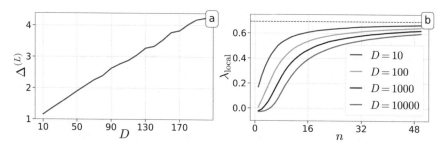

**Fig. 12.1 a** Fractal dimension (estimated via the Lyapunov dimension, Sect. 5.9) of the logistic transport map (12.1). **b** Instantaneous exponential growth rates of infinitesimal random perturbations averaged all over the state space vs. time $n$. The horizontal dashed line gives the value log(2) of the largest Lyapunov exponent of the system

## 12.2  Tipping Points in Dynamical Systems

In the following sections we now switch gears and go from the realistic view of climate as a complex system with billions of degrees of freedom, to a conceptual representation much more like that of Sect. 2.1. Such representations can be useful when one cares about the climate state that is averaged over long time periods or long spatial scales (e.g., global mean temperature).

Here we will discuss what happens to the long term evolution of a dynamical system under external perturbations of the state or parameters, a sufficiently slow parameter variation, or noise input. While most of the time such input is reflected as a mild response from the system, here we will discuss cases where external input causes *transitions* from one dynamic regime (e.g., a stable fixed point) to a sufficiently different one (e.g., a different stable fixed point). Because such a transition would never occur if the dynamical system was evolving as-is, without the external input, we will call them *induced transitions*. Such induced transitions are useful to study in systems that display *multi-stability*, or equivalently, *coexisting attractors*.

Alternatively, we can be talking in terms of *tipping points* (or *regime shifts* or *critical transitions*), which is the terminology used in recent literature to describe the same phenomena. Given the size (or duration) of an impulse, the system may *tip* from one stable regime to another. The tipping points do not necessarily have to be "points", i.e., singular instances of parameter combinations, but can also be intervals or ranges of value combinations that lead to tipping, so it is useful to think more in terms of "tipping scenarios".

### 12.2.1  Tipping Mechanisms

An induced transition (or tipping) can have significantly different origins, so in the following we will try to do a classification. Regardless of the mechanism, in every case there are two practically interesting quantities. First is the *tipping threshold*. This means, how large should the magnitude of the external perturbation be for tipping to be triggered. For example, in a case of multi-stability as in the magnetic pendulum of Fig. 5.4, the magnitude must necessarily be at least as large as the typical radius of the basin of attraction around each magnet. The second quantity is the *typical tipping timescale*. That is, provided that we have confirmed that the external perturbation has sufficient magnitude, how much time will the system need to tip from one regime to the other? Answering either of these questions can only be done in a context-specific manner.

With this said, let's now go through some typical tipping mechanisms. Illustrations for the mechanisms are provided in Fig. 12.2. The list we will present here is definitely incomplete, as many kinds of mechanisms can induce transitions across attracting states and many more will probably be uncovered in the near future.

**Noise-driven.** This scenario is probably the simplest one conceptually.[1] It targets cases where a system may transition from one attractor to another because of noise input perturbing the state of the system. If the maximum magnitude of the noise perturbation exceeds the typical radius of the basin of attraction, then the system will be driven out of it and start converging towards another attractor. A simple example is shown in Fig. 12.2a with cyan color (the state can only go up and down as $p$ remains fixed in this scenario; the reason the cyan line also goes left-right a bit is only for visually illustrating something "noisy").

For an example based on a real dynamical system we can visit again Fig. 5.4. The system state there is residing in one of the three fixed points and to it we apply a noisy perturbation. As long as the amplitude of the noise is small, the state will hover over the vicinity of the same fixed point. However, if the noise amplitude is too large, then the state will be brought into the fractal part of the basins of attraction, and could just as easily converge to any of the other two attractive fixed points.

**Bifurcation-based.** Imagine a dynamical system with a parameter $p$ that has two attractors $A_1(p)$, $A_2(p)$ each with its own basin of attraction $B_1(p)$, $B_2(p)$. The system is originally "resting" on $A_1$. The word resting can be taken literally if $A_1$ is a stable fixed point, or a bit more figuratively if $A_1$ is a periodic or chaotic attractor. Then, a system parameter $p$ is varied infinitely slowly: slow enough that the system equilibration to the slightly varied attractor $A_1(p)$ is instant w.r.t. the parameter variation rate. At some critical value $p_c$ a bifurcation occurs and $A_1$ loses its stability after the bifurcation, and either becomes unstable or simply does not exist anymore. In this case the system tips to $A_2$ which still exists. Importantly, reversing the parameter change does not lead the system back to the original state even though

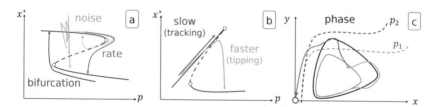

**Fig. 12.2** Sketches illustrating mechanisms leading to induced transitions. **a, b** are bifurcation diagrams, with solid black curves denoting stable, and dashed curves denoting unstable fixed points. On top of them different mechanisms are illustrated schematically. **c** is a state space plot of a 2D dynamical system. The solid triangle-like circles are stable limit cycles and the dashed curves separate the basins of attraction of the limit cycles from those of the stable fixed point at the origin. The color denotes their location for each parameter $p_1$, $p_2$. Orange and cyan are two perturbations of the state originally on the purple limit cycle, via a quick successive change of the parameter $p_1 \rightarrow p_2 \rightarrow p_1$

---

[1] To be fair, to analyze and detect noise driven tipping in detail in a continuous time system, one would need to utilize stochastic differential equations, but we chose to not include this topic in our textbook.

$A_1$ is once again existing in the state space. That is because the system state is now trapped into the basin of attraction of $A_2$.

This is illustrated in Fig. 12.2a (purple line), and we have already seen this scenario in Sect. 4.1.1, where we discussed the concept of hysteresis using an exemplary simple 1D energy balance model. While bifurcation-based induced transitions are typically presented with both attractors $A_1$, $A_2$ being fixed points, any other attracting set works here as well.

**Rate-dependent.** This mechanism targets cases where whether tipping will occur or not depends on how fast (i.e., the rate) a parameter is changed. There are two different conceptual scenarios that we can consider to understand this. The first scenario is based on the preceding example of two coexisting attractors $A_1$ and $A_2$ with bifurcation parameter $p_c$, see Fig. 12.2a. The parameter $p$ is varied from $p_1 < p_c$ to $p_2 > p_c$ and back to $p_1$ in a given time span $\Delta t$. This change can be smooth or instantaneous at two time points $t_1$, $t_2$, and these details don't matter for the concept. Crucially, the transition phase of the dynamical system towards the new attractor takes finite time. Depending on how small $\Delta t$ is, or how large the original distance $|p_2 - p_1|$ is, the system may be able to re-enter the basin of attraction of $A_1$ or not as shown schematically in Fig. 12.2a with dotted orange (no transition) and solid orange (transition) color.

Let's make the picture concrete by using the simplified 1D climate model of (2.2). We examine the scenario where $\epsilon$ is forced back and forth with an expression like $\epsilon(t) = \epsilon_0 + \Delta\epsilon \exp\left(-\frac{(t-t_0)^2}{2\Delta t^2}\right)$. We write down the equations in an expression that keeps time explicitly, as

$$\frac{dT}{dt} = 0.5 + 0.2 \tanh\left(\frac{T - 263}{4}\right) - \epsilon \cdot 10 \cdot (0.002T)^4$$

$$\frac{d\epsilon}{dt} = -(t - t_0)\frac{\Delta\epsilon}{\Delta t^2} \exp\left(-\frac{(t - t_0)^2}{2\Delta t^2}\right). \tag{12.2}$$

The result of evolving this system for 3 combinations of $(\Delta\epsilon, \Delta t)$ is shown in Fig. 12.3. As we can see the system may, or may not transition from the nearest (upper) stable state to one in the further away (lower) stable branch of temperatures, depending on $\Delta\epsilon$, $\Delta t$.

Another case that depends on the rate of change of $p$ is shown in Fig. 12.2b. In this scenario we don't change and revert $p_1 \rightarrow p_2 \rightarrow p_1$ but instead monotonically change $p$ with finite speed. If we change $p$ too fast, the system might not have time to equilibrate to the new fixed point (or in general attractor) and the trajectory crosses a very close-by basin of attraction of another attractor. If the change of $p$ is sufficiently slow, the system always equilibrates to the nearest fixed point, or "tracks" it.

**Phase-dependent.** This kind of tipping is prominent in dynamical systems where limit cycles are common, for example predator-pray models. Let's have a look at Fig. 12.2c which is the sketch of the state space of a 2D typical predator-prey dynamical system (see Further reading). The system has an attracting limit cycle and a stable

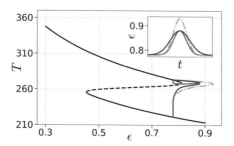

**Fig. 12.3** Illustration of (12.2). The main plot, which was perceived as a bifurcation diagram in Fig. 4.1, can now be seen as a state space plot if one considers $\epsilon$ a dynamic variable. The colored lines show the evolution of the system, starting from $(T_0, \epsilon_0) = (278, 0.78)$. Each color has different time forcing given by $(\Delta\epsilon, \Delta t) = (0.1, 40), (0.15, 40), (0.1, 60)$, as also shown in the inset

fixed point at the origin. The limit cycle exist for parameter value $p_1$ (purple) and $p_2$ (black) but slightly changes in structure and size. The dashed lines illustrate the boundary between the basin of attraction of the limit cycle and the fixed point for each parameter.

Now, imagine the system state oscillating in the purple limit cycle at parameter $p_1$, and a short impulse changes $p_1$ to $p_2$ and back again. In the cyan case, the trajectory goes quickly towards the adjusted limit cycle, and then returns to the original one. However, if we perform the same impulse at a different *phase* of the oscillation, something else happens. In the orange trajectory the system again goes quickly to the adjusted limit cycle. However, now when $p_2 \to p_1$ the system will not go back to the original limit cycle, but instead to the fixed point at the origin. That is because the state found itself in the basin of attraction of the fixed point instead. Have a look online at animations/12/phase_dependent for an interactive demonstration.

## 12.2.2 Basin Stability and Resilience

It is useful to be able to quantify in some sense whether a dynamical system is tending towards a tipping point. While a lot of information on this topic will be given in Further reading, here we will focus on a specific global quantifier. In the literature global quantifiers of stability are often referred to as *resilience*.

So far in this book when discussing stability we have used ideas that look at a small neighborhood around a state space point, and can thus be labelled *local*. In Chap. 1 we labelled fixed points as stable or unstable using the eigenvalues of the Jacobian at the fixed point. Using the Jacobian, which is the linearized form of the dynamic rule $f$, is only valid for small perturbations, which is why it restricts us locally around the fixed point of interest. In Chap. 3 we used the Lyapunov exponents to characterize chaotic motion. However, the LEs are also based on local concepts (the Jacobian). In Chap. 4 we looked mostly at local bifurcations with only a short discussion on global bifurcations.

The limitation of local concepts is that they can say very little about the basin of attraction $\mathcal{B}_A$ of an attractor $A$, because $\mathcal{B}_A$ is generated necessarily by the nonlinear, or *global*, dynamics. What becomes more and more apparent in this section, is that we care not only whether some fixed points become stable or unstable (the standard perspective of local bifurcation analysis), but also how their basin of attraction changes with the change of a parameter. Furthermore, the significance of a non-local (or equivalently non-small) perturbation depends on whether the perturbation will bring the system state outside $\mathcal{B}_A$ or not, which once again cannot be quantified by local measures.

A simple global quantifier of the stability of a basin of attraction is its state space volume fraction $F(\mathcal{B}_A)$. For dynamical systems that do not have a finite state space (e.g., $\mathcal{S} = \mathbb{R}^D$) a context-specific physically sensible finite state space region should be used instead of $\mathcal{S}$. One then calculates $F(\mathcal{B}_A)$, which coincides with the probability that a state space point initialized randomly in $\mathcal{S}$ will be in $\mathcal{B}_A$. For low dimensional dynamical systems $F(\mathcal{B}_A)$ can be calculated by initializing a grid-based covering of initial conditions on the state space, and recording the fraction of those that converge to $A$. For higher dimensional systems this quickly becomes unfeasible to compute and a better alternative is to initialize $M$ states uniformly randomly in $\mathcal{S}$ and record the fraction $F(\mathcal{B}_A)$ of them that converges to $A$. The error of the true estimate with this approach converges to 0 with $M \to \infty$ as $\sqrt{F(\mathcal{B}_A)(1 - F(\mathcal{B}_A))/M}$. Using DynamicalSystems.jl this can be done with the function `basin_fractions`.

To apply the concept to a concrete example, let's modify the magnetic pendulum of Sect. 5.3.3 as

$$\dot{\mathbf{x}} = \mathbf{v}, \quad \dot{\mathbf{v}} = -\omega^2 \mathbf{x} - q\mathbf{v} - \sum_{i=1}^{N} \frac{\gamma_i (\mathbf{x} - \mathbf{x}_i)}{d_i^3}. \tag{12.3}$$

with $\mathbf{x} = (x, y)$ and $\mathbf{v} = (v_x, v_y)$. The only difference with Sect. 5.3.3 is that now each magnet has a different magnetic field strength $\gamma_i$. The system has three fixed points with velocities 0 and the magnet positions $\mathbf{x}_i$. The (local) stability of the fixed points can be determined by calculating the eigenvalues of the Jacobian matrices there, and specifically recording the maximum real value $\lambda_1$ of the eigenvalues as discussed in Chap. 1.

In Fig. 12.4 we show how the basin fraction changes with $\gamma_3$, i.e., the basin stability[2] measure $F$ and $\lambda_1$ for one of the three points while varying $\gamma_3$. The local measure $\lambda_1$ remains almost entirely constant and negative, and only when $\gamma_3 \approx 0$ it changes value to a positive one. On the other hand $F$ continuously decreases to 0 as we decrease $\gamma_3$, being a clear indicator that the fixed point loses stability on a global level as we decrease $\gamma_3$. Do note, however, that the basin stability measure has a large

---

[2] Quick disclaimer: we cheated when making Fig. 12.4, because we only used initial conditions with $\mathbf{v}_0 = 0$, for visual purposes. In practice one has to cover the entire state space with initial conditions.

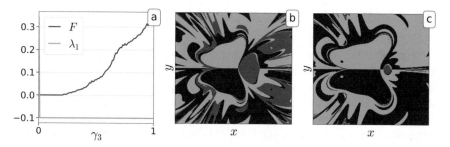

**Fig. 12.4  a** Basin stability $F$ and largest real part of the Jacobian eigenvalues at the fixed point corresponding to the third magnet of the system of (12.3) versus $\gamma_3$, with $\gamma_1 = \gamma_2 = 1$. **b, c** Basins of attraction for $\gamma_3 = 0.7$ and $\gamma_3 = 0.25$. showing the subset of the state space that is used for computing the basin stability $F$

downside: it cannot be applied straightforwardly to real world data, since knowledge of the dynamic rule $f$ is necessary to calculate it.

### 12.2.3  Tipping Probabilities

A concept directly related with the above one are *tipping probabilities*. They aim to quantify what is the probability to tip to a specific attractor, given a parameter change and the attractor basin overlap before and after the change. Let $\mathcal{B}_i(p)$ denote the basin of attraction of attractor $A_i$ at parameter(s) $p$. We define the tipping probability from $A_i$ to $A_j$, given an instantaneous parameter change $p_- \rightarrow p_+$, as

$$P(A_i \rightarrow A_j | p_- \rightarrow p_+) = \frac{|\mathcal{B}_j(p_+) \cap \mathcal{B}_i(p_-)|}{|\mathcal{B}_i(p_-)|} \qquad (12.4)$$

where $|\cdot|$ is simply the volume of the enclosed set. The equation describes something quite simple: what is the overlap of the basin of attraction of $A_i$ at $p_-$ with that of the attractor $A_j$ at $p_+$. The formula is still valid if $A_i$ does not exist at $p_+$ or if $A_j$ does not exist at $p_-$, a property that comes up as useful in many cases. The parameter change $p_- \rightarrow p_+$ can be just as well replaced by a temporal change $t_- \rightarrow t_+$ in the case of time forced non-autonomous systems. Like in Sect. 12.2.2, for infinite state spaces we consider a finite subset.

Notice that (12.4) has two downsides: (1) probabilities are estimated only based on initial and final states, which neglects the "journey" in between. In cases where the attractor basins are smooth this is okay, but if one has fractal boundaries then $P$ will depend sensitively on the choice of p- and/or $p_+$. (2) The formula does not take into account the natural density (or measure, Sect. 8.3). If the "initial state" at $p_-$ is a trajectory already converged to an attractor, then it is more important to quantify the overlap of the attractor with the new basin at $p_+$. See exercises for this version.

## 12.3   Nonlinear Dynamics Applications in Climate

We've reached the very last section of this book (congratulations by the way if you made it all the way through!). Here we want to do something a little bit different. Instead of teaching a useful new concept or a practical consideration, we will showcase applications of the topics you've just learned in this book, to study climate, or climate-related aspects, as a dynamical system. This is partly because these applications are just so fascinating, we couldn't resist. But also because we want to showcase how nonlinear dynamics can be extremely useful in understanding dynamical aspects of climate, and motivate usage of simple models more in this field. Think of the remaining of this section as an "extended further reading".

### 12.3.1   Excitable Carbon Cycle and Extinction Events

The carbon cycle is the cycling of carbon atoms between their main reservoirs: biosphere (living matter), lithosphere (soil, sediments, fossils, . . .), hydrosphere (oceans, . . .) and atmosphere. It can be studied using data from ocean sediments or other geochemical records. Disruptions of the cycle, due to, e.g., massive volcanic activity or bolide impacts or other "forcings", can be seen as relatively abrupt and large changes in the isotopic composition of sedimentary carbon. The largest of these disruptions are associated with extinction events: periods of Earth's past characterized by sharp decreases in biodiversity and biomass [275–277].

The typical interpretation of the impact of forcings on the recorded carbon data is that the magnitude of the carbon fluctuations is proportional to the magnitude of the forcing, and the (sedimentary part of) the carbon cycle passively records the forcing. Here we will present the work of Rothman [278], who argues that the dynamics of the carbon cycle are better represented by a low dimensional excitable system, surprisingly similar to the FitzHugh-Nagumo model of Sect. 2.2.4, rather than a linear response.

In Rothman's model the upper ocean is a "box" with an influx of dissolved calcium carbonate from rivers and outflux of carbon into the deep ocean and sediments. Within the ocean inorganic carbon reactions take place, converting $CO_2$ to $HCO_3^-$ to $CO_3^{2-}$ and back. The concentrations of the various forms of carbon (including organic) react to imbalances in the fluxes and in the biological consumption and production of $CO_2$. This provides means of amplifying small perturbations (i.e., *excitability*). We will skip the remaining details necessary to formulate the model and refer to [278], and see also [279] for a mathematical introduction to the dynamics of the carbon cycle. The final model reads

$$\dot{c}/f(c) = \mu[1 - bs(c, c_p) - \theta\bar{s}(c, c_x) - \nu] + w - w_0$$
$$\dot{w} = \mu[1 - bs(c, c_p) + \theta\bar{s}(c, c_x) - \nu] - w + w_0 \qquad (12.5)$$

**Fig. 12.5** An excitable carbon cycle due to Rothman [278] showing the effect of an additional injection of dissolved inorganic carbon in the upper ocean of strength $v$ which excites the system for $v \geq 0.39$

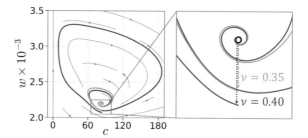

with $c$ the $CO_3^{2-}$ concentration and $w$ the concentration of all inorganic carbon, $CO_2 + HCO_3^- + CO_3^{2-}$ . All other terms are parameters and functions related to the dynamics of the carbon cycle, with $s, \bar{s}, f$ being nonlinear functions of $c$, e.g., $s(c, c_p) = c^\gamma / (c^\gamma + c_p^\gamma)$, see code online for exact values. The only parameter we will care about here is $v$, which represents the extra injection of $CO_2$ from external forcings like volcanoes or humans. The impact of $v$ in this model is almost identical to the impact of $I$ in (2.6). A finite-time impulse with $v > 0$ may excite the system or not, as shown in Fig. 12.5.

In [278, 280] Rothman discusses how disturbances of the carbon cycle in geochemical timeseries can be viewed in this model. Minor disruption events correspond to perturbations slightly below the threshold that triggers excitation. Major disturbances, associated with massive extinction events, correspond to perturbations well above the threshold. For more about this fascinating story, see [278–284]. Furthermore, excitable systems have been used in climate dynamics in other contexts as well, e.g., glaciation cycles [285, 286] and peatlands [287].

### 12.3.2 Climate Attractors

Can the dynamics of climate be represented by a low-dimensional chaotic attractor, similar to that of the Lorenz-63 system? Initially it might sound absurd to even ask such a question, given that climate is a complex system. Nevertheless as the theory of chaos and nonlinear dynamics started gaining popularity within the scientific community around the 1980s, this was one of the first questions asked about measured climate data.

Starting at 1984, Nicolis and Nicolis [288] used some of the earliest paleoclimate measurements (oxygen isotope records from Pacific deep sea drilled cores) and delay coordinates embedding (Chap. 6) to "reconstruct" a potential climate attractor represented by the data. They calculated a fractal dimension for these data using the correlation sum (Chap. 5), and found it to be about 3.1, saturating already after a $d = 4$ dimensional embedding. Many subsequent studies by Fraedrich and many others [289–297] analyzed other various forms of climatic timeseries and, via the same methods, obtained fractal dimensions in the range of 3–8.

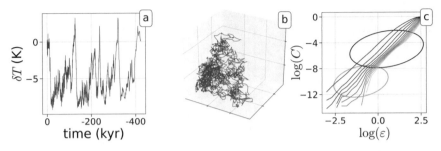

**Fig. 12.6** Estimating the fractal dimension of the "attractor" underlined by the Vostok Ice Core data. **a** Original timeseries, interpolated linearly on an equi-spaced time axis. **b** Delay reconstruction with appropriate delay time. **c** Correlation sum versus $\varepsilon$ for embeddings of **a** with $d$ ranging from 3 (dark blue) to 12 (light yellow)

Such small values for the fractal dimension of such a complex system naturally raised a lot of skepticism. In fact, some pioneers of nonlinear dynamics, Grassberger and Procaccia (who developed the correlation sum algorithm to estimated the fractal dimension) and Lorenz (see further reading of Chap. 3) were openly skeptical of such estimates [298–301], showing by several means how these estimates might not be as accurate as presented. Given what you've read so far in Chap. 5, you can see how the data used in these estimates fall into one, or several, common "pitfalls" of estimating the fractal dimension: they are not stationary, they are too few, they are noisy, they are not equi-temporally sampled, they are not even sampled but rather represent time-averages, or they are interpolated.

Let us highlight the dangers of applying the concepts of this book in real world data without care. We will do a similar kind of analysis using data from the Vostok Ice Core [302], shown in Fig. 12.6. We computed the correlation sum for these data over the range of all reasonable $\varepsilon$ values. If one attempted to estimate the fractal dimension $\Delta$ from Fig. 12.6c, by estimating the "slope" of the entire curve via linear regression, one would get $\Delta \approx 2.7$ saturating quite rapidly for increasing $d$. There are several problems however. First, as indicated by the two ellipses in Fig. 12.6c, the curves are separated into two fundamentally different regimes. In the first (cyan) there is a very high slope that does not saturate versus $d$ and the lines are also not straight. The other regime (black) has smaller slope, but once again the curve $\log(C)$ versus $\log(\varepsilon)$ is not a straight line, which makes estimating a fractal dimension from it invalid.

### 12.3.3   Glaciation Cycles as a Driven Oscillator Problem

As is clearly evident in Fig. 12.6a, the history of Earth's climate reveals that it has gone through several temperature "oscillations" between very cold, or glacial, states and relatively warm, or interglacial, states with a characteristic period of approxi-

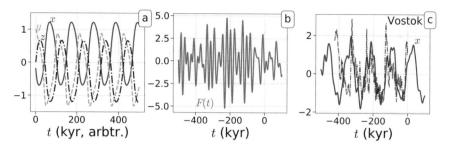

**Fig. 12.7  a** Limit cycle of model (12.6) for parameters $p = 1.0, q = 2.5, r = 1.3, s = 0.6, u = 0.2$ and no forcing, $v = 0$. **b** Astronomical forcing, as in [286]. **c** $x$ timeseries of the forced van der Pol oscillator, (12.7) in purple and Vostoc ice core data over-plotted in black. For **b, c** the 0 of the time axis is approximately the present

mately 100 kyr. This well-established observational fact is most typically referred to as *glaciation cycles*, or *ice-ages*, which have occurred over the Pleistocene period and have typical timescales of 10–100 kyr. A sensible theoretical approach to approach this problem is the so-called Milankovich theory, proposed first by Milankovich in 1984 [303]. The premise of the theory is straightforward to comprehend: long-timescale variations in Earth's orbital parameters, also called *astronomical forcing*, cause the total insolation reaching the regions of high latitudes to vary in a quasiperiodic manner. Decrease in insolation triggers an increase of ice, leading to a glaciation state, and vice versa for the warm state.

Discussing the successes and shortcomings of this theory is a book in itself, and we cannot go that deep here. However, we can present some exemplary work that applies the concepts of forced dynamical systems and synchronization we discussed in Chap. 9. One of the most important contributions to this approach was by Saltzman and coworkers, who established a theory in which glaciation cycles are interpreted as a limit cycle synchronized on the astronomical forcing [304, 305]. They've created a physically inspired model that connects the long-term variations of continental ice mass $x$, $CO_2$ concentration $y$ and deep ocean temperature $z$. The model, presented here already in de-dimensionalized form, is [306]

$$\dot{x} = -x - y - uz - vF(t), \quad \dot{y} = -pz + ry - sy^2 - y^3, \quad \dot{z} = -q(x + z)$$
$$(12.6)$$

with $u, v, p, r, s, q$ positive parameters and $F(t)$ representing the astronomical forcing, normalized to unit variance. $F$ is typically assumed quasiperiodic, with most power in periods 19, 23 ky (precession) and 41 ky (obliquity). Equations (12.6) can already display free oscillations (i.e., a limit cycle) in the unforced scenario of $v = 0$ with period close to the period of the observations, as shown in Fig. 12.7a. Forcing the system however with, e.g., $v = 0.5$ results in an impressive match between the produced oscillations and the observed paleoclimate data (see Fig. 5 of [306]). One of the conclusions of that work is that ice ages would occur today even in the absence of astronomical forcing due the natural internal nonlinear dynamics of climate. The main effect of the astronomical forcing is to control the timing of glaciations.

Now, model (12.6) is great and physically inspired, but one can go even simpler. Due to the universality of limit cycles and synchronization in dynamical systems (Chap. 9), it is likely that any model with an internal periodicity of 100 kyr could be used to model the glaciation cycles. Enter our good old friend the van der Pol oscillator, which was used by Crucifix and co-workers to prove exactly that [286]. Let

$$\tau \dot{x} = -y + \beta + \gamma F(t), \quad \tau \dot{y} = -\alpha(y^3/3 - y - x) \tag{12.7}$$

with $F$ once again the astronomical forcing and $\tau$ a necessary parameter to tune the period of the limit cycle to an appropriate value (these equations are equivalent with the ones in (9.3)). Here $x$, the slow variable of the system, corresponds to ice volume while $y$, the fast variable, can be interpreted somewhat at liberty, e.g., as the Atlantic ocean circulation. As we know, the free van der Pol system obtains a limit cycle, while with forcing it may perform chaotic oscillations with period similar to that of the limit cycle. Indeed, we use $\alpha = 11.11$, $\beta = 0.25$, $\gamma = 0.75$, $\tau = 40.09$ to tune the system into a limit cycle of approximately 100 kyr ($t$ is measured in kyr in (12.7)). We then force the system with forcing $F(t)$ shown in Fig. 12.7b. In Fig. 12.7c we show the resulting $x$, the slow variable of the system which corresponds to ice volume. On top we plot the same Vostok data from Fig. 12.6 without any modification. We also took little care to "optimize" model parameters and initial condition for a good fit, but nevertheless we see sensible results overall. A review of using oscillating dynamical systems to model the glaciation cycles is given by Crucifix in [285] and goes into more depth in this fascinating story.

## Further Reading

Two noteworthy and general textbooks that connect nonlinear dynamics with climate are *Nonlinear Climate Dynamics* by Dijkstra [307] and *Mathematics and Climate* by Kaper and Engler [308].

An excellent review text on the predictability of complex systems, on which Sect. 12.1 is based on, is Chap. 8 of [51] by Siegert and Kantz. See also *Predictability of Weather and Climate* by Palmer et al. [309] for more in depth discussion specifically on weather. Prediction of systems with multiple scales has been discussed many times by Lorenz [102, 310], who played a pioneering role on the topic. His Lorenz-96 model [102] is a prototypical and simple enough dynamical system that is used to study multi-scale dynamics. It has been recently extended by Thornes, Düben and Palmer [311] to have an even larger scale hierarchy and was used to discuss scale-dependent precision of modelling and forecasting. Equally interesting is the more recent publication [312] by Brisch and Kantz, which discusses how strictly finite prediction horizons can arise in nonlinear systems with multiple temporal and spatial scales, and is also based on an extension of the Lorenz-96 model.

Induced transitions (tipping points) are important for many areas of science, not only climate change. Other prominent examples are ecology and population dynamics [313–317] and disease spreading in human populations [318–320]. Some general references on critical tipping elements of the climate system are [284, 321–325] among others. A textbook devoted to bifurcations and catastrophes (older terminology of tipping points) was written by Arnold et al. [326]. The classification of tipping points based on bifurcations, noise, or rate, was done first by Ashwin et al. [327], and here we simply expanded the list with a few more "versions" of tipping from recent literature. A recent discussion connecting tipping points with invariant measures (Chap. 8) in climate can be found in [328]. Phase-dependent induced transitions were introduced recently by Alkhayuon et al. [329]. The same reference also provides a mathematical framework to estimate how sensitive such a transition is. Tipping points are inherently linked with multi-stability, a concept with much larger history in dynamical systems theory. To keep the discussion here short, we point out to recent reviews by Pisarchik and Feudel [330, 331] and references therein.

Naturally, effort has been put to devise methods that could potentially detect critical transitions *before* they occur. These are typically called *early warning indicators*, see Scheffer et al. [332] and Brovkin et al. [333]. For example in [334] Dakos et al. show a timeseries-based approach that can identify tipping points based on the slowing down of dynamics. Kuehn [335] extended theory to handle systems with two timescales and see also the work of Thompson and Sieber [336] for another predictive approach and [336] for another review on noise-based approaches. Ritche and Sieber discuss indicators for rate-induced tipping in [337]. A second review on this topic can be found in [323] which puts the ideas more into the context of societal systems. It should be noted that many of these early warning indicators that rely on slowing down of the dynamics have found only limited success in practical applications, as many systems that undergo tipping do not actually show any slowing down [338, 339].

Basin stability, as defined in Sect. 12.2.2, was introduced recently by Menck et al. [339]. For a recent application in a more complicated dynamical system that couples several potential tipping elements of Earth see [340]. The related concept of *tipping probabilities* was discussed in an article by Kaszás, Feudel and Tél [341]. The same reference discusses in a more concrete setting tipping in chaotic systems which (typically) have fractal basins of attraction.

In our book we decided to focus on deterministic dynamical systems exclusively. Stochastic or random dynamical systems (which are then typically represented by stochastic differential equations, SDEs) have been employed numerous times to study the dynamics of climate as well as tipping points in general, given how important stochastic driving is there. For such aspects we point to an extensive recent review by Ghil and Lucarini [342] (which also contains more information related to this chapter) and references therein. Ghil has also done a lot of work in simplified dynamical systems approaches to climate, particularly in energy balance models, see, e.g., [343]. Another stochasticity-related concept with relevance for climate and several other scientific areas is that of the *stochastic resonance*, introduced first by Benzi Sutera and Vulpiani [344, 345]. It also exceeds scope here, so we refer to reviews on it [346, 347].

## Selected Exercises

12.1 Write a function that calculates the instantaneous growth rate of perturbations $\lambda_{local}$, just like in Fig. 12.1b, but do not average the growth rate over the state space. Color-code the attractor of the Hénon map (3.3) with the value of $\lambda_{local}$ as calculated for each starting position on the attractor. *Hint: to calculate $\lambda_{local}$ use (3.1) for short times, and then average over all possible perturbation orientations.*

12.2 Give a theoretical justification, using the Jacobian, of exactly what the state space plot of $\lambda_{local}$ of the previous exercise should look like, when computed for a single step.

12.3 Repeat the first exercise for the chaotic attractor of the Lorenz-63 system, (1.5). Calculate the local growth rates for a short time, e.g., $\Delta t = 1$. You will notice a characteristic region of the attractor with high $\lambda_{local}$. This indicates an unstable periodic orbit (many such orbits are "hidden" inside the chaotic attractor).

12.4 A simple model that can showcase rate-dependent tipping is Stommel's box model for Atlantic thermohaline circulation [348, 349] given by

$$\dot{T} = \eta_1 - T - |T - S| \cdot T, \quad \dot{S} = \eta_2 - \eta_3 S - |T - S| \cdot S.$$

Here $T, S$ are variables standing for dimensionless temperature and salinity differences between the boxes (polar and equatorial ocean basins), respectively, and $\eta_i$ are parameters. For the exercise keep $\eta_2 = 1, \eta_3 = 0.3$ and produce a bifurcation diagram for $T$ versus $\eta_1$ from 2 to 4, which should look very similar to Fig. 12.2b.

12.5 Continue from the above exercise and do a simulation where $\eta_1$ is increasing from 2.5 to 3.33 with a constant rate $d\eta$ during the time evolution (start your simulation from the fixed point!). For each $d\eta$ find at which $\eta_1$ value the system will tip from the initial stable branch to the second one, by crossing the unstable fixed point branch. *Hint: At $\eta_1 \approx 3.33$ the original fixed point branch gets annihilated and we have a bifurcation-based tipping. We are interested in tipping that happens before this 3.33 value, because this indicates that the system can no longer track the original attractor and undergoes a rate-dependent tipping as in Fig. 12.2b!.*

12.6 An example of tipping scenarios in ecology can be found in the Rosenzweig-MacArthur predator-prey model [350]. It describes the connection between a fast-evolving prey population $x$ and a slowly-reproducing predator population $y$. The model equations in non-dimensionalized form following [351] are

$$\kappa \dot{x} = x(1 - \phi x) - \frac{xy}{1 + \eta x}, \quad \dot{y} = \frac{xy}{1 + \eta x} - y. \tag{12.8}$$

Here $\kappa \ll 1$ is the timescale separation between the reproduction rates of the prey and the predator, $\phi$ is inversely proportional to the carrying capacity of the prey, and $\eta$ can be interpreted as the predator's handling time of the prey. Use $\kappa = 0.01, \eta = 0.8$ in the following and $\phi$ will stay bounded in the range $\phi \in (\phi_-, \phi_+) = (\eta(1 - \eta)(1 + \eta), 1 - \eta)$. Find the fixed points of the system and describe their linear stability within this parameter range. Now modify the system so that $\phi$ is increasing monotonously with time with constant rate, $\dot{\phi} = r$, starting from $\phi_-$ until $\phi_+$ where it stays constant there (this is similar to (12.2) but with a much simpler temporal dependence). Start your simulations always from the fixed point with $x$-coordinate $x^* = 1/(1 - \eta)$. As $\phi$ increases this fixed point remains linearly stable. However, just like in Fig. 12.2b, if the rate $r$ is high enough, the system fails to track the fixed point and instead undergoes a critical transition to a *temporary* collapse of the prey population $x$. Find the $r_c$ where for $r > r_c$ we have this temporary collapse. *Hint: use $r \in (0, 0.01)$ and analyze the full timeseries evolution in search of temporary $x$ collapse, because the final point of the timeseries is the same regardless if $r > r_c$ or not.*

12.7 Consider the Lorenz-84 model

$$\dot{x} = -y^2 - z^2 - ax + aR, \quad \dot{y} = xy - y - bxz + G, \quad \dot{z} = bxy + xz - z$$

$$(12.9)$$

with parameters $R = 6.886, G = 1.347, a = 0.255, b = 4.0$. For these parameters in the state space there are three coexisting attractors: a fixed point, a periodic trajectory, and a chaotic attractor. Devise a scheme that labels each initial condition according to which attractor it ends up at. Use this to calculate their basin fractions $F(\mathcal{B})$ with the random sampling method of $N = 10,000$ initial conditions discussed in Sect. 12.2.2. *Hint: there are many approaches to label the initial conditions. You could use their spectral properties, their Lyapunov spectrum, or their fractal dimension, or you could devise your own method (in each case always first evolve the initial conditions for some long transient). If the code takes too long, reduce $N$ to 1,000.*

12.8 Continue the above exercise and calculate the basin stability of the fixed point attractor versus decreasing the parameter $G$ to 1.27 (i.e., make a figure like Fig. 12.4a).

12.9 Repeat the above exercise using two different methods of labelling initial conditions. Comment on the differences between the three methods regarding (i) computational performance, (ii) robustness (how much the fractions $F$ fluctuate for a different sample of $N$ random initial conditions), and (iii) which method requires the least amount of tweaking by hand the parameters of the labelling algorithm.

12.10 Consider the tipping probabilities defined in Sect. 12.2.3. Draw a state space sketch of a dynamical system with 3 attractors at $p_-$ and $p_+$, where their basins of attraction change shape and size but none vanish. Illustrate the sets used in (12.4) in a Venn diagram fashion. Based on your sketch argue why it should be that $\sum_j P(A_i \to A_j | p_- \to p_+) = 1$. The summation happens over all attractors that exist at parameter $p_+$, which may, or may not, include $i$.

12.11 Apply the tipping probabilities concept to the magnetic pendulum discussed in Fig. 12.4. Calculate the tipping probability from magnet $i$ to magnet $j$ where $p_-$ corresponds to $\gamma_3 = 1$ and $p_+$ to $\gamma_3 = 0.4$. Do the probabilities you find abide the system's symmetries?

12.12 Continue from the above exercise, and focus on the tipping probability from magnet 3 to 2. Calculate $P$ versus $\gamma_3$ for values ranging from 1 to 0. Before you run the code, what value do you expect for $P$ when $\gamma_3 \rightarrow 0$? Can you confirm your expectation?

12.13 Equation (12.4) does not target tipping scenarios where the current state of the system is already on an attractor. Provide an alternative, but similar definition that instead takes into account the natural density of the attractor and its overlap with basins of attraction of other attractors that exist after the parameter change $p_- \rightarrow p_+$. *Hint: it is important that you "weight" the basins of attraction by the natural density so that basins that are in higher density of the attractor have higher probability of tipping.*

# Appendix A
# Computing Lyapunov Exponents

In Sect. 3.2.2 we presented an algorithm for estimating the Lyapunov spectrum of a dynamical system. This, however, was provided with an ad hoc argument of equating the diagonal entries of the $R$ matrix (of the QR-decomposition) with the Lyapunov exponents. Here, we revisit this algorithm and explicitly justify where this relationship comes from.

Recall from Sect. 1.5 that the Jacobian of the dynamical rule $f$ provides the rate of increase (for discrete systems the direct increase) of infinitesimal perturbations around a specific point in the state space $\mathbf{x}$ and the Lyapunov exponents quantify the average stretching (or contracting) of perturbations in characteristic directions on the invariant set.

To compute the evolution of an infinitesimal perturbation $\mathbf{y}_0$ applied to the initial state $\mathbf{x}_0$ of a continuous system we consider the evolution of the perturbed state $\mathbf{x}_0 + \mathbf{y}_0$ and truncate the Taylor expansion of $f$, $\dot{\mathbf{x}} + \dot{\mathbf{y}} = f(\mathbf{x} + \mathbf{y}) = f(\mathbf{x}) + J_f(\mathbf{x}) \cdot \mathbf{y}$, such that

$$\dot{\mathbf{y}} = J_f(\mathbf{x}) \cdot \mathbf{y} \qquad (A.1)$$

where $\mathbf{x}$ denotes a solution of $\dot{\mathbf{x}} = f(\mathbf{x})$ given by the flow, $\mathbf{x}(t) = \Phi^t(\mathbf{x}_0)$. Formally, the evolution of $\mathbf{y}(t)$ is given by the Jacobian matrix of the flow $J_{\Phi^t}(\mathbf{x}_0)$

$$\mathbf{y}(t) = J_{\Phi^t}(\mathbf{x}_0) \cdot \mathbf{y}_0 \qquad (A.2)$$

because $\mathbf{x}(t) + \mathbf{y}(t) = \Phi^t(\mathbf{x}_0 + \mathbf{y}_0) = \Phi^t(\mathbf{x}_0) + J_{\Phi^t}(\mathbf{x}_0) \cdot \mathbf{y}_0$. The evolution of $k$ different perturbations can be compactly described by considering them as column vectors of a $(D, k)$ matrix $Y$ and extending (A.1) to a matrix ODE $\dot{Y} = J_f(\mathbf{x}) \cdot Y$ with initial values $Y(0) = Y_0$. Using (A.2) the evolution of this matrix $Y(t)$ can also be expressed using the Jacobian of the flow, $Y(t) = J_{\phi^t}(\mathbf{x}_0) \cdot Y_0$. If the Jacobian matrix of the flow is needed explicitly you can compute it using the $(D, D)$ identity matrix as initial condition $Y_0$ in the matrix ODE for $Y$. Note, however, that in the following we will use $J_{\Phi^t}(\mathbf{x}_0)$ only to explain the theoretical frame work of the algorithm for computing Lyapunov exponents. Practically, the evolution of all perturbations,

© The Editor(s) (if applicable) and The Author(s), under exclusive license to Springer Nature Switzerland AG 2022
G. Datseris and U. Parlitz, *Nonlinear Dynamics*, Undergraduate Lecture Notes in Physics, https://doi.org/10.1007/978-3-030-91032-7

being vectors in the tangent space of the state space, is computed using ODEs like
(A.1) involving the Jacobian matrix $J_f(\mathbf{x})$ of the vector field $f$ along the trajectory
$\mathbf{x}(t) = \Phi^t(\mathbf{x}_0)$, only.

In a similar way one can derive an iteration rule for the dynamics of perturbations
in discrete systems and the computation of the linearized dynamics can thus be
summarized as

$$\begin{aligned} \dot{\mathbf{x}} &= f(\mathbf{x}) \\ \dot{Y} &= J_f(\mathbf{x}) \cdot Y \end{aligned} \quad \text{or} \quad \begin{aligned} \mathbf{x}_{n+1} &= f(\mathbf{x}_n) \\ Y_{n+1} &= J_f(\mathbf{x}_n) \cdot Y_n \end{aligned} \tag{A.3}$$

starting at $\mathbf{x}_0$ with initial perturbations $Y_0$ and evolving $Y(t)$ in parallel with the
state $\mathbf{x}$ so that the state dependent Jacobian $J_f(\mathbf{x}(t))$ is used at each point along the
trajectory.

As pointed out in Sect. 3.2.2 Lyapunov exponents quantify the *average* stretching
rates on an invariant set. To get a good covering of the set one therefore has to simulate
(A.3) for a long time $T$. The computation of the Jacobian matrix of the flow $J_{\Phi^T}(\mathbf{x}_0)$
for large $T$ suffers, however, from severe numerical issues due over- or underflow (as
a result of exponentially increasing or decreasing matrix elements) and the fact that all
columns of this matrix will converge towards the most quickly expanding direction.
Although you initialize (A.3) for computing the Jacobian matrix $J_{\Phi^T}(\mathbf{x}_0)$ with the
identity matrix the column vectors of $J_{\Phi^t}(\mathbf{x}_0)$ (or $Y(t)$) will not remain orthogonal
during their evolution but converge to the first Lyapunov vector. Thus, after some time
(depending on the difference between $\lambda_1$ and the remaining exponents) your solution
for $J_{\Phi^t}(\mathbf{x}_0)$ is effectively useless, because all its column vectors have become (almost)
colinear. To avoid this problem you have to split the long simulation interval $[0, T]$
into shorter time intervals and perform some renormalization and reorthogonalization
of solution vectors after each time interval.

We will now use the above considerations to come up with an algorithm that allows
computing the $k \le D$ largest Lyapunov exponents, by studying the growth rate of
the (infinitesimally small) $k$-dimensional sub-volume $V_k$. $V_k(t)$ will exponentially
increase (or decrease) as $V_k(t) \approx V_k(0) \exp(\Lambda_k t)$ with[1] $\Lambda_k = \sum_{j=1}^{k} \lambda_j$. If the volume
growth rates $\Lambda_k$ are known for different $k$ we can compute the Lyapunov exponents
as: $\lambda_1 = \Lambda_1$, $\lambda_2 = \Lambda_2 - \lambda_1 = \Lambda_2 - \Lambda_1$, and in general $\lambda_j = \Lambda_j - \Lambda_{j-1}$ for $j = 1, \ldots, k$.

So in essence we need to compute the volume growth rates $\Lambda_k$ and this provides
all Lyapunov exponents $\lambda_j$ with $j \le k$. Let's return to the $D \times k$ matrix $Y(t)$. To
compute the volume $V_k(t)$ of the parallelepiped spanned by the column vectors of
$Y$ we decompose this matrix into the product $Y = Q \cdot R$ of an orthonormal $(D, k)$
matrix $Q$ and a $(k, k)$ upper triangular matrix $R$ with non-negative diagonal elements.[2] The volume $V_k(t)$ is then given by the product of the diagonal elements of

---

[1] Because the volume of a $k$-dimensional ellipsoid is $V(t) = \frac{4}{3}\pi \prod_{i=1}^{k}(\epsilon \exp(\lambda_i t))$, it increases
exponentially in time with exponent $\Lambda_k = \sum_{i=1}^{k} \lambda_i$.

[2] The $QR$-decomposition $Y = Q \cdot R$ is a *unique* factorization, that can be computed via Gram–
Schmidt reorthogonalization or Householder transformations. The columns of $Q$ are orthonormal,
i.e., $Q^{tr}Q = I_k$ where $I_k$ is the $(k, k)$ identity matrix.

**Fig. A.1** Volume $V_2 = R_{11} R_{22}$ of the parallelepiped spanned by two column vectors $\mathbf{y}^1$ and $\mathbf{y}^2$ of the matrix $Y = Q \cdot R$. The vectors $\mathbf{q}^1$ and $\mathbf{q}^2$ are the column vectors of $Q$, respectively, with $\|\mathbf{q}^i\| = 1$, and $R_{ij}$ are elements of the upper-triangular matrix $R$

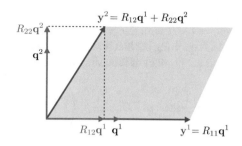

$R(t)$

$$V_k(t) = \prod_{i=1}^{k} R_{ii}(t) \tag{A.4}$$

as illustrated for $D = 2$ in Fig. A.1.

Using an orthonormal set of perturbation vectors for initialization we get[3] $V_k(0) = 1$ and thus

$$\Lambda_k t \approx \ln(V_k(t)) = \ln\left(\prod_{i=1}^{k} R_{ii}(t)\right) = \sum_{i=1}^{k} \ln(R_{ii}(t)) \tag{A.5}$$

and so

$$\lambda_j = \Lambda_j - \Lambda_{j-1} \approx \frac{1}{t} \sum_{i=1}^{j} \ln(R_{ii}(t)) - \frac{1}{t} \sum_{i=1}^{j-1} \ln(R_{ii}(t)) = \frac{1}{t} \ln(R_{jj}(t)). \tag{A.6}$$

These approximative results for $\Lambda_j$ and $\lambda_j$ become exact in the limit $t \to \infty$.

Remember, our goal is to establish a computational scheme based on small time steps combined with renormalization and reorthogonalization to avoid any numerical issues. To do so we use the chain rule to decompose the Jacobian of the flow for the full time interval $[0, T]$ into a product of Jacobian matrices for $N$ small time steps $\Delta t = T/N$ and perform repeated $QR$-decomposition for each step

$$
\begin{aligned}
Q(T) \cdot R(T) = Y(T) &= J_{\Phi^T}(\mathbf{x}_0) \cdot Y(0) \\
&= J_{\Phi^{\Delta t}}(\Phi^{(N-1)\Delta t}(\mathbf{x}_0)) \cdot \ldots \cdot J_{\Phi^{\Delta t}}(\Phi^{\Delta t}(\mathbf{x}_0)) \cdot \underbrace{J_{\Phi^{\Delta t}}(\mathbf{x}_0) \cdot Y(0)} \\
&= J_{\Phi^{\Delta t}}(\Phi^{(N-1)\Delta t}(\mathbf{x}_0)) \cdot \ldots \cdot J_{\Phi^{\Delta t}}(\Phi^{\Delta t}(\mathbf{x}_0)) \cdot Q(\Delta t) \cdot R(\Delta t) \\
&= J_{\Phi^{\Delta t}}(\Phi^{(N-1)\Delta t}(\mathbf{x}_0)) \cdot \ldots \cdot Q(2\Delta t) \cdot R(2\Delta t) \cdot R(\Delta t) \\
&= Q(N\Delta t) \cdot R(N\Delta t) \cdot \ldots \cdot R(\Delta t).
\end{aligned} \tag{A.7}
$$

---

[3] You may have a concern now, since a volume of "1" is not at all infinitesimal. But $Y$ is evolved using the *linearized* dynamics that "does not care about" the absolute value of the perturbations.

Now since the $QR$-decomposition is unique we can equate the $R$-matrix on the left side of (A.7) with the product of $R$-matrices computed along the orbit on the right side. Exploiting the fact that the diagonal elements of a product of triangular matrices equal the product of their diagonal elements we finally obtain

$$\lambda_j \approx \frac{1}{T} \ln(R_{jj}(T)) = \frac{1}{T} \ln\left(\prod_{n=1}^{N} R_{jj}(n\Delta t)\right) = \frac{1}{N\Delta t} \sum_{n=1}^{N} \ln\left(R_{jj}(n\Delta t)\right). \quad \text{(A.8)}$$

To summarize, we can compute the average stretching rates over a long time interval $[0, T]$ by evolving the orthogonal $D \times k$ matrices $Q(n\Delta t)$ for (sufficiently) short periods of time $\Delta t$ along the orbit and perform $QR$-decompositions of the resulting matrices. Formally, the evolution of $Q(n\Delta t)$ is given by mapping it with the Jacobian of the flow, $J_{\Phi^{\Delta t}}(\Phi^{n\Delta t}(\mathbf{x}_0)) \cdot Q(n\Delta t)$. Practically, however, you can perform this step by solving (A.3) with starting $Y(0) = Q(n\Delta t)$, i.e., you never have to explicitly compute the Jacobian $J_{\Phi^{\Delta t}}$, you only need $J_f$ in (A.3), but this matrix is in general known analytically or can be computed using automatic differentiation of $f$.

For continuous systems it is also possible to derive differential equations for the matrices $Q$ and $R$ and use their solutions for computing Lyapunov exponents. Inserting $Y(t) = Q(t)R(t)$ in the matrix ODE $\dot{Y} = J_f(\mathbf{x})Y$ yields $\dot{Q}R + Q\dot{R} = J_f(\mathbf{x})QR$ and multiplying by the inverse $R^{-1}$ and exploiting $Q^{tr}Q = I_k$ results in

$$W = \dot{R}R^{-1} = Q^{tr}J_f(\mathbf{x})Q - S \quad \text{(A.9)}$$

with $S = Q^{tr}\dot{Q}$. $Q^{tr}Q = I_k$ implies $\dot{Q}^{tr}Q + Q^{tr}\dot{Q} = 0$, i.e., $S = Q^{tr}\dot{Q} = -\dot{Q}^{tr}Q = -(Q^{tr}\dot{Q})^{tr} = -S$ which means that $S$ is skew-symmetric. Since the inverse of an upper triangular matrix is also upper triangular and the product of two upper-triangular matrices is again upper-triangular, the matrix $W$ is also upper-triangular, such that skewness of $S$ and (A.9) imply

$$S_{ij} = \begin{cases} (Q^{tr}J_f(\mathbf{x})Q)_{ij} & \forall j < i \\ 0 & i = j \\ -(Q^{tr}J_f(\mathbf{x})Q)_{ji} & \forall j > i. \end{cases} \quad \text{(A.10)}$$

The desired differential equation for the matrix $Q$ can thus be written as $\dot{Q} = J_f(\mathbf{x})Q - Q\dot{R}R^{-1} = J_f(\mathbf{x})Q - Q(Q^{tr}J_f(\mathbf{x})Q - S)$ or

$$\dot{Q} = J_f(\mathbf{x})Q - QQ^{tr}J_f(\mathbf{x})Q + QS \quad \text{(A.11)}$$

which for $k = D$ simplifies to $\dot{Q} = QS$, because in this case $QQ^{tr} = I_D$. For computing the Lyapunov exponents we have to quantify the growth rate the logarithms $\rho_i = \log(R_{ii})$ of the diagonal elements. Taking the derivative yields

$$\dot{\rho}_i = \frac{\dot{R}_{ii}}{R_{ii}} = W_{ii} = (Q^{tr} J(\mathbf{x}) Q)_{ii} \tag{A.12}$$

and the Lyapunov exponent $\lambda_i$ is thus given by $\rho_i / t$ for $t \to \infty$ or

$$\lambda_i = \lim_{t \to \infty} \frac{1}{t} \int_0^t (Q^{tr} J_f(\mathbf{x}) Q)_{ii} dt \tag{A.13}$$

that can be interpreted as averaging of the values of the local Lyapunov exponent $\lambda_i^{\text{loc}}(\mathbf{x}) = (Q^{tr} J_f(\mathbf{x}) Q)_{ii}$ along the trajectory $\mathbf{x} = \mathbf{x}(t)$. Unfortunately, practically using this formalism for computing Lyapunov exponents is a challenge because you have to make sure that the matrix $Q$ remains orthonormal when solving the corresponding ODE. But we can use this framework to prove for $k = D$ the evolution equation $\dot{v} = \text{trace}(J_f(\mathbf{x}(t)) \cdot v$ for an infinitesimal $D$-dimensional volume $v = V_D$ that was introduced in Sect. 1.5 to characterize conservative and dissipative continuous systems. This volume is given as $v = \det(Y) = \prod_{i=1}^D R_{ii}$ and its derivative reads $\dot{v} = \sum_{i=1}^D \frac{\dot{R}_{ii}}{R_{ii}} \prod_{j=1}^D R_{jj} = v \sum_{i=1}^D \frac{\dot{R}_{ii}}{R_{ii}} = v \, \text{trace}(W) = v \, \text{trace}(J_f(\mathbf{x}(t)))$, because $\text{trace}(Q^{tr} J_f(\mathbf{x}) Q) = \text{trace}(J(\mathbf{x}))$ (for $k = D$ we have $Q^{tr} = Q^{-1}$).

An equation for the volume growth rate $\Lambda_D = \lim_{t \to \infty} \frac{1}{t} \log(v(t)/v(0))$ can be derived using the logarithm of the volume $z(t) = \log(v(t)/v(0))$. The evolution equation of $z$ is given by $\dot{z} = \dot{v}/v = \text{trace}(J_f(\mathbf{x}(t)))$, i.e., $z(t) = \int_0^t \text{trace}(J_f(\mathbf{x}(t')))dt'$. The $D$-dimensional volume growths rate thus equals the temporal average of the trace of the Jacobian matrix of $f$

$$\Lambda_D = \lim_{t \to \infty} \frac{1}{t} \int_0^t \text{trace}(J_f(\mathbf{x}(t')))dt' = \langle \text{trace}(J_f(\mathbf{x}) \rangle_t. \tag{A.14}$$

$\Lambda_D = \lambda_1 + \cdots + \lambda_D$ is negative for dissipative dynamics and vanishes for trajectories following a conservative time evolution ($\to$ Liouville's theorem).

# Appendix B
# Deriving the Master Stability Function

We start the discussion with (10.1) from Chap. 10

$$\dot{\mathbf{x}}_i = g(\mathbf{x}_i) - \sigma \sum_{j=1}^{N} G_{ij} h(\mathbf{x}_j) \tag{B.1}$$

where $\sigma$ is a global parameter for controlling the coupling strength and $G$ is a $(N, N)$ coupling matrix given by $G_{ij} = -W_{ij}$ for $i \neq j$ and $G_{ii} = \sum_{j=1}^{N} W_{ij}$. $W$ is the weighted adjacency matrix of the network with $W_{ii} =$). The zero-row condition for $G$ implies that the fully synchronized state $\mathbf{x}_1(t) = \mathbf{x}_2(t) = \cdots = \mathbf{x}_N(t) = \mathbf{s}(t)$ is a solution of this network. This condition defines a $d$ dimensional linear subspace of the $D = Nd$ dimensional state space of the network called the *synchronization manifold*.

To perform a linear stability analysis of this synchronous solution we consider an infinitesimally small perturbation $\mathbf{y}(t)$ of $\mathbf{s}(t)$ and linearize the equations of motion at $\mathbf{s}(t)$. Using the network equations in (B.1) the linearized equations describe the evolution of perturbations $\mathbf{y}_i$ for each node as

$$\dot{\mathbf{y}}_i = \left[ J_g(\mathbf{s}) - \sigma \sum_{j=1}^{N} G_{ij} J_h(\mathbf{s}) \right] \mathbf{y}_i. \tag{B.2}$$

with $J_g, J_h$ the Jacobian matrices of $g, h$, respectively. Using the direct product $\otimes$ of matrices (also called Kronecker product) this set of linearized equations may be combined

$$\dot{\mathbf{y}} = \left[ I_N \otimes J_g(\mathbf{s}) - \sigma G \otimes J_h(\mathbf{s}) \right] \mathbf{y} \tag{B.3}$$

where $\mathbf{y} = (\mathbf{y}_1, \mathbf{y}_2, \ldots, \mathbf{y}_N) \in \mathbb{R}^D$ represents the perturbation in the full state space. We will use (B.3) to investigate the dynamics of the perturbation of the synchronized state.

© The Editor(s) (if applicable) and The Author(s), under exclusive license to Springer Nature Switzerland AG 2022
G. Datseris and U. Parlitz, *Nonlinear Dynamics*, Undergraduate Lecture Notes in Physics, https://doi.org/10.1007/978-3-030-91032-7

To further simplify the stability analysis we assume that the coupling between nodes $i$ and $j$ is symmetric, i.e., $G_{ij} = G_{ji}$. This implies that $G$ has only real eigenvalues $\mu_k$ and can be diagonalized $E = (\mu_1, \mu_2, \ldots, \mu_N) = U^{-1}GU$ by means of an orthogonal $(N, N)$ matrix $U$. The matrix $U$ can also be used to introduce a new coordinate system for the perturbations with some block structure that facilitates the stability analysis. This transformation is defined as $\mathbf{z} = (U^{-1} \otimes I_d)\mathbf{y}$ where $I_d$ denotes the $(d, d)$ identity matrix. To derive an equation for the inverse transformation and further computations the following two identities for general (invertible) matrices $A, B, C, D$ will be employed

$$(A \otimes B)^{-1} = A^{-1} \otimes B^{-1} \tag{B.4}$$

$$AC \otimes BD = (A \otimes B)(C \otimes D). \tag{B.5}$$

Using (B.4) we get $\mathbf{y} = (U \otimes I_d)\mathbf{z}$. Substituting in (B.3) $\mathbf{y}$ by $\mathbf{z}$ and applying (B.5) the linearized equations in the new coordinate system are given by

$$\begin{aligned}
\dot{\mathbf{z}} &= (U^{-1} \otimes I_d) \left[ I_N \otimes J_g(\mathbf{s}) - \sigma G \otimes J_h(\mathbf{s}) \right] (U \otimes I_d)\mathbf{z} \\
&= \left[ U^{-1}U \otimes J_g I_d - \sigma U^{-1}GU \otimes J_h I_d \right] \mathbf{z} \\
&= \left[ I_N \otimes J_g(\mathbf{s}) - \sigma E \otimes J_h \right] \mathbf{z}.
\end{aligned} \tag{B.6}$$

Since $I_N$ and $E$ are both diagonal matrices this set of ODEs consists of $N$ $(d, d)$ blocks (or subsystems) that do not interact and can therefore be analyzed separately. The dynamics of the $k$th block is characterized by the $k$th eigenvalue $\mu_k$ of the coupling matrix $G$

$$\dot{\mathbf{z}}_k = \left[ J_g(\mathbf{s}) - \sigma \mu_k J_h(\mathbf{s}) \right] \mathbf{z}_k. \tag{B.7}$$

Since all these equations have the same form

$$\dot{\mathbf{z}}_k = \left[ J_g(\mathbf{s}) - \alpha J_h(\mathbf{s}) \right] \mathbf{z}_k \tag{B.8}$$

it suffices to investigate the stability of (B.8) by computing the largest Lyapunov exponent $\lambda_1$ as a function of the parameter $\alpha$. This function $\lambda_1(\alpha)$ is called *master stability function*.

Master stability functions can also be defined for more general cases. For example, if the coupling matrix is not symmetric but diagonalizable, then the dynamics of (B.8) has to be investigated for complex $\alpha$. Furthermore, extensions of the approach for networks consisting of almost identical systems and for networks of networks have been devised (but are beyond the scope of our presentation here).

# References

1. J. Gleick, *Chaos: Making a New Science* (Viking Books, New York, 1987)
2. P. Holmes, History of dynamical systems. Scholarpedia **2**(5), 1843 (2007). Revision #91357
3. F. Diacu, P. Holmes, *Celestial Encounters: the Origins of Chaos and Stability*, vol. 22 (Princeton University Press, Princeton, 1999)
4. S.H. Strogatz, *Nonlinear Dynamics and Chaos: with Applications to Physics, Biology, Chemistry, and Engineering (Studies in Nonlinearity)* (CRC Press, Boca Raton, 2000)
5. J. Argyris, G. Faust, M. Haase, R. Friedrich, *An Exploration of Dynamical Systems and Chaos* (Springer, Berlin, 2015)
6. R.M. May, Simple mathematical models with very complicated dynamics. Nature **261**(5560), 459–467 (1976)
7. B.V. Chirikov, A universal instability of many-dimensional oscillator systems. Phys. Rep. **52**(5), 263–379 (1979)
8. E.N. Lorenz, Deterministic nonperiodic flow. J. Atmos. Sci. **20**(2), 130–141 (1963)
9. M. Henon, C. Heiles, The applicability of the third integral of motion: some numerical experiments. Astron. J. **69**, 73 (1964)
10. C. Gissinger, A new deterministic model for chaotic reversals. Eur. Phys. J. B **85**(4), 137 (2012)
11. R. Anderson, R. May, Population biology of infectious diseases: part i. Nature **280**, 361–367 (1979)
12. W.O. Kermack, A.G. McKendrick, A contribution to the mathematical theory of epidemics. Proc. R. Soc. Lond. A **115**, 700–721 (1927)
13. I. Bendixson, Sur les courbes définies par des équations différentielles. Acta Math. **24**, 1–88 (1901)
14. M.I. Budyko, The effect of solar radiation variations on the climate of the Earth. Tellus **21**(5), 611–619 (1969)
15. W.D. Sellers, Global climatic model based on the energy balance of the Earth-atmosphere system (1969)
16. A.M. Zhabotinsky, Belousov-Zhabotinsky reaction. Scholarpedia **2**(9), 1435 (2007). Revision #91050
17. A.T. Winfree, The prehistory of the Belousov-Zhabotinsky oscillator. J. Chem. Educ. **61**(8), 661 (1984)
18. I. Prigogine, R. Lefever, Symmetry breaking instabilities in dissipative systems. ii. J. Chem. Phys. **48**(4), 1695–1700 (1968)
19. B. Kralemann, L. Cimponeriu, M. Rosenblum, A. Pikovsky, R. Mrowka, Uncovering interaction of coupled oscillators from data. Phys. Rev. E **76**, 055201 (2007)

G. Datseris and U. Parlitz, *Nonlinear Dynamics*, Undergraduate Lecture Notes in Physics,
https://doi.org/10.1007/978-3-030-91032-7

20. R. FitzHugh, Mathematical models of threshold phenomena in the nerve membrane. Bull. Math. Biophys. **17**, 257–278 (1955)
21. R. FitzHugh, Impulses and physiological states in theoretical models of nerve membrane. Biophys. J. **1**(6), 445–466 (1961)
22. J. Nagumo, S. Arimoto, S. Yoshizawa, An active pulse transmission line simulating nerve axon. Proc. IRE **50**(10), 2061–2070 (1962)
23. R. Fitzhugh, Motion picture of nerve impulse propagation using computer animation. J. Appl. Physiol. **25**(5), 628–630 (1968). PMID: 5687371
24. E.M. Izhikevich, R. FitzHugh, FitzHugh-Nagumo model. Scholarpedia **1**(9), 1349 (2006). Revision #123664
25. A.S. Pikovsky, J. Kurths, Coherence resonance in a noise-driven excitable system. Phys. Rev. Lett. **78**, 775–778 (1997)
26. B. Lindner, J. García-Ojalvo, A. Neiman, L. Schimansky-Geier, Effects of noise in excitable systems. Phys. Rep. **392**(6), 321–424 (2004)
27. E.M. Izhikevich, *Dynamical Systems in Neuroscience: the Geometry of Excitability and Bursting* (MIT Press, Cambridge, 2006)
28. E. Ott, *Chaos in Dynamical Systems* (Cambridge University Press, Cambridge, 2012)
29. F. Hoppensteadt, Predator-prey model. Scholarpedia **1**(10), 1563 (2006). Revision #91667
30. J.D. Murray (ed.), *Mathematical Biology* (Springer, New York, 2002)
31. W.G. Hoover, Remark on "some simple chaotic flows". Phys. Rev. E **51**, 759–760 (1995)
32. Wikipedia contributors, For want of a nail — Wikipedia, the free encyclopedia (2019). Accessed 30 Oct 2019
33. H. Poincaré, Sur le problème des trois corps et les équations de la dynamique. Acta Math. **13**(1), A3–A270 (1890)
34. J. Barrow-Green, *Poincare and the Three Body Problem*. History of Mathematics, vol. 11 (American Mathematical Society, Providence, 1996)
35. E. Ott, Edward N. Lorenz (1917–2008). Nature **453**(7193), 300–300 (2008)
36. R. Abraham, C.D. Shaw, *Dynamics: the Geometry of Behavior. Part 2: Chaotic Behavior*. Visual Mathematics Library: Vismath Books (Aerial Press, Santa Cruz, 1985)
37. A.M. Lyapunov, The general problem of the stability of motion. Int. J. Control **55**(3), 531–534 (1992)
38. A. Politi, Lyapunov exponent. Scholarpedia **8**(3), 2722 (2013). Revision #137286
39. V. Oseledets, Oseledets theorem. Scholarpedia **3**(1), 1846 (2008). Revision #142085
40. I. Shimada, T. Nagashima, A numerical approach to ergodic problem of dissipative dynamical systems. Prog. Theor. Phys. **61**(6), 1605–1616 (1979)
41. G. Benettin, L. Galgani, A. Giorgilli, J.-M. Strelcyn, Lyapunov characteristic exponents for smooth dynamical systems and for Hamiltonian systems; a method for computing all of them, part 1: theory. Meccanica **15**(1), 9–20 (1980)
42. A. Wolf, J.B. Swift, H.L. Swinney, J.A. Vastano, Determining Lyapunov exponents from a time series. Phys. D: Nonlinear Phenom. **16**(3), 285–317 (1985)
43. K. Geist, U. Parlitz, W. Lauterborn, Comparison of different methods for computing Lyapunov exponents. Prog. Theor. Phys. **83**(5), 875–893 (1990)
44. Ch. Skokos, The Lyapunov characteristic exponents and their computation. Lect. Notes Phys. **790**, 63–135 (2010)
45. A. Pikovsky, A. Politi, *Lyapunov Exponents* (Cambridge University Press, Cambridge, 2016)
46. B.R. Hunt, E. Ott, Defining chaos. Chaos **25**(9), 097618 (2015)
47. Ch. Skokos, T.C. Bountis, Ch. Antonopoulos, Geometrical properties of local dynamics in Hamiltonian systems: the generalized alignment index (GALI) method. Phys. D: Nonlinear Phenom. **231**(1), 30–54 (2007)
48. M. Fouchard, E. Lega, C. Froeschlé, C. Froeschlé, On the relationship between fast Lyapunov indicator and periodic orbits for continuous flows. Celest. Mech. Dyn. Astron. **83**(1), 205–222 (2002)
49. R. Barrio, W. Borczyk, S. Breiter, Spurious structures in chaos indicators maps. Chaos Solitons Fractals **40**(4), 1697–1714 (2009)

50. P.M. Cincotta, C.M. Giordano, C. Simó, Phase space structure of multi-dimensional systems by means of the mean exponential growth factor of nearby orbits. Phys. D: Nonlinear Phenom. **182**(3), 151–178 (2003)
51. C. Skokos, G.A. Gottwald, J. Laskar (eds.), *Chaos Detection and Predictability*. Lecture Notes in Physics, vol. 915 (Springer, Berlin, 2016)
52. N.P. Maffione, L.A. Darriba, P.M. Cincotta, C.M. Giordano, A comparison of different indicators of chaos based on the deviation vectors: application to symplectic mappings. Celest. Mech. Dyn. Astron. **111**(3), 285 (2011)
53. H. Wernecke, B. Sándor, C. Gros, How to test for partially predictable chaos. Sci. Rep. **7**(1), 1087 (2017)
54. V. Baran, A.A. Raduta, Classical and quantal chaos described by a fourth order quadrupole boson Hamiltonian. Int. J. Mod. Phys. E **07**(04), 527–551 (1998)
55. J.C. Sprott, A dynamical system with a strange attractor and invariant tori. Phys. Lett. A **378**(20), 1361–1363 (2014)
56. Yu.A. Kuznetsov, *Elements of Applied Bifurcation Theory*, 3rd edn. (Springer, Berlin, 2004)
57. M. Haragus, G. Iooss, *Local Bifurcations, Center Manifolds, and Normal Forms in Infinite-Dimensional Dynamical Systems* (Springer, London, 2011)
58. E.L. Allgower, K. Georg, *Numerical Continuation Methods* (Springer, Berlin, 1990)
59. H. Dankowicz, F. Schilder, *Recipes for Continuation* (Society for Industrial and Applied Mathematics, Philadelphia, 2013)
60. M.J. Feigenbaum, Quantitative universality for a class of nonlinear transformations. J. Stat. Phys. **19**(1), 25–52 (1978)
61. P. Coullet, C. Tresser, Itérations d'endomorphismes et groupe de renormalisation. J. Phys. Colloq. **39**(C5), C5–25–C5–28 (1978)
62. S. Grossmann, S. Thomae, Invariant distributions and stationary correlation functions of one-dimensional discrete processes. Z. Naturforsch. A **32**(12), 1353–1363 (1977)
63. C. Tresser, P. Coullet, E. de Faria, Period doubling. Scholarpedia **9**(6), 3958 (2014). Revision #142883
64. W. Lauterborn, E. Cramer, Subharmonic route to chaos observed in acoustics. Phys. Rev. Lett. **47**, 1445–1448 (1981)
65. P. Cvitanović, *Universality in Chaos: a Reprint Selection* (Adam Hilger, Bristol, 1989)
66. Y. Pomeau, P. Manneville, Intermittent transition to turbulence in dissipative dynamical systems. Commun. Math. Phys. **74**(2), 189–197 (1980)
67. P. Manneville, Intermittency, self-similarity and 1/f spectrum in dissipative dynamical systems. J. Phys. **41**(11), 1235–1243 (1980)
68. G. Datseris, L. Hupe, R. Fleischmann, Estimating Lyapunov exponents in billiards. Chaos: Interdiscip. J. Nonlinear Sci. **29**(9), 093115 (2019)
69. H.G. Schuster, W. Just, *Deterministic Chaos* (Wiley-VCH Verlag GmbH & Co. KGaA, Weinheim, 2005)
70. K.A. Maasch, B. Saltzman, A low-order dynamical model of global climatic variability over the full Pleistocene. J. Geophys. Res. **95**(D2), 1955 (1990)
71. C.E. Shannon, A mathematical theory of communication. Bell Syst. Tech. J. **27**(3), 379–423 (1948)
72. C.E. Shannon, W. Weaver, *The Mathematical Theory of Communication* (University of Illinois Press, Urbana, 1949)
73. A. Rényi, On the dimension and entropy of probability distributions. Acta Math. Acad. Sci. Hung. **10**(1–2), 193–215 (1959)
74. T.M. Cover, J.A. Thomas, *Elements of Information Theory*. Wiley Series in Telecommunications and Signal Processing, 2nd edn. (Wiley, New York, 2006)
75. M. Ribeiro, T. Henriques, L. Castro, A. Souto, L. Antunes, C. Costa-Santos, A. Teixeira, The entropy universe. Entropy **23**(2), 1–35 (2021)
76. H. Kantz, T. Schreiber, *Nonlinear Time Series Analysis*, 2nd edn. (Cambridge University Press, Cambridge, 2003)

77. T. Tel, M. Gruiz, *Chaotic Dynamics, an Introduction Based on Classical Mechanics* (Cambridge University Press, Cambridge, 2007)
78. K. Falconer, *Fractal Geometry: Mathematical Foundations and Applications* (Wiley, New York, 2013)
79. B.B. Mandelbrot, W.H. Freeman, and Company, *The Fractal Geometry of Nature*. Einaudi Paperbacks (Henry Holt and Company, New York, 1983)
80. F. Hausdorff, Dimension und äusseres Mass. Math. Ann. **79**(1–2), 157–179 (1918)
81. V.S. Anishchenko, G.I. Strelkova, Irregular attractors. Discret. Dyn. Nat. Soc. **2**(1), 53–72 (1998)
82. J. Theiler, Estimating fractal dimension. J. Opt. Soc. Am. A **7**(6), 1055 (1990)
83. G. Datseris, I. Kottlarz, A.P. Braun, U. Parlitz, Estimating the fractal dimension: a comparative review and open source implementations (2021) arxiv: 2109.05937
84. L. Pontrjagin, L. Schnirelmann, Sur une propriete metrique de la dimension. Ann. Math. **33**(1), 156 (1932)
85. K. Falconer, *Fractal Geometry Mathematical Foundations and Applications* (Wiley, New York, 2003)
86. D.A. Russell, J.D. Hanson, E. Ott, Dimension of strange attractors. Phys. Rev. Lett. **45**(14), 1175–1178 (1980)
87. P. Grassberger, I. Procaccia, Characterization of strange attractors. Phys. Rev. Lett. **50**(5), 346–349 (1983)
88. P. Grassberger, I. Procaccia, Measuring the strangeness of strange attractors. Phys. D: Nonlinear Phenom. **9**(1–2), 189–208 (1983)
89. P. Frederickson, J.L. Kaplan, E.D. Yorke, J.A. Yorke, The Lyapunov dimension of strange attractors. J. Differ. Equ. **49**(2), 185–207 (1983)
90. F. Takens, On the numerical determination of the dimension of an attractor, in *Dynamical Systems and Bifurcations*, ed. by B.L.J. Braaksma, H.W. Broer, F. Takens (Springer, Berlin, 1985), pp. 99–106
91. G. Paladin, A. Vulpiani, Anomalous scaling laws in multifractal objects. Phys. Rep. **156**(4), 147–225 (1987)
92. D. Ruelle, *Chaotic Evolution and Strange Attractors* (Cambridge University Press, Cambridge, 1989)
93. E. Ott, Attractor dimensions. Scholarpedia **3**(3), 2110 (2008). Revision #91015
94. J.-P. Eckmann, D. Ruelle, Fundamental limitations for estimating dimensions and Lyapunov exponents in dynamical systems. Phys. D: Nonlinear Phenom. **56**(2), 185–187 (1992)
95. P. Grassberger, I. Procaccia, Characterization of strange attractors. Phys. Rev. Lett. **50**, 346–349 (1983)
96. J. Theiler, Spurious dimension from correlation algorithms applied to limited time-series data. Phys. Rev. A **34**(3), 2427–2432 (1986)
97. Wikipedia contributors, Kaplan-Yorke conjecture — Wikipedia, the free encyclopedia (2020). Accessed 12 June 2020
98. J.L. Kaplan, J.A. Yorke, Chaotic behavior of multidimensional difference equations, in *Functional Differential Equations and Approximation of Fixed Points*, ed. by H.-O. Peitgen, H.-O. Walther (Springer, Berlin, 1979), pp. 204–227
99. C. Grebogi, S.W. McDonald, E. Ott, J.A. Yorke, Final state sensitivity: an obstruction to predictability. Phys. Lett. A **99**(9), 415–418 (1983)
100. S.W. McDonald, C. Grebogi, E. Ott, J.A. Yorke, Fractal basin boundaries. Phys. D: Nonlinear Phenom. **17**(2), 125–153 (1985)
101. E. Rosenberg, *Generalized Dimensions and Multifractals* (Springer International Publishing, Cham, 2020), pp. 325–364
102. E.N. Lorenz, Predictability: a problem partly solved, in *Seminar on Predictability, 4-8 September 1995*, vol. 1 (Shinfield Park, Reading, 1995), pp. 1–18. ECMWF
103. H.D.I. Abarbanel, *Analysis of Observed Chaotic Data* (Springer, New York, 1996)
104. E. Bradley, H. Kantz, Nonlinear time-series analysis revisited. Chaos **25**(9), 097610 (2015)

105. N.H. Packard, J.P. Crutchfield, J.D. Farmer, R.S. Shaw, Geometry from a time series. Phys. Rev. Lett. **45**(9), 712–716 (1980)
106. J.P. Eckmann, D. Ruelle, Fundamental limitations for estimating dimensions and Lyapunov exponents in dynamical systems. Phys. D: Nonlinear Phenom. **56**(2–3), 185–187 (1992)
107. F. Takens, Detecting strange attractors in turbulence, in *Dynamical Systems and Turbulence, Warwick 1980*, ed. by D. Rand, L.-S. Young (Springer, Berlin, 1981), pp. 366–381
108. H. Whitney, Differentiable manifolds. Ann. Math. **37**(3), 645–680 (1936)
109. T. Sauer, J.A. Yorke, M. Casdagli, Embedology. J. Stat. Phys. **65**(3–4), 579–616 (1991)
110. T. Sauer, J.A. Yorke, How many delay coordinates do you need? Int. J. Bifurc. Chaos **03**(03), 737–744 (1993)
111. M. Casdagli, S. Eubank, J.D. Farmer, J. Gibson, State space reconstruction in the presence of noise. Phys. D: Nonlinear Phenom. **51**(1), 52–98 (1991)
112. T. Sauer, Reconstruction of dynamical systems from interspike intervals. Phys. Rev. Lett. **72**(24), 3811–3814 (1994)
113. J. Stark, Delay embeddings for forced systems. I. Deterministic forcing. J. Nonlinear Sci. **9**(3), 255–332 (1999)
114. J. Stark, D.S. Broomhead, M.E. Davies, J. Huke, Delay embeddings for forced systems. II. Stochastic forcing. J. Nonlinear Sci. **13**(6), 519–577 (2003)
115. C. Letellier, L.A. Aguirre, Investigating nonlinear dynamics from time series: the influence of symmetries and the choice of observables. Chaos: Interdiscip. J. Nonlinear Sci. **12**(3), 549–558 (2002)
116. C. Letellier, J. Maquet, L. Le Sceller, G. Gouesbet, L.A. Aguirre, On the non-equivalence of observables in phase-space reconstructions from recorded time series. J. Phys. A: Math. Gen. **31**(39), 7913–7927 (1998)
117. D.A. Smirnov, B.P. Bezruchko, Y.P. Seleznev, Choice of dynamical variables for global reconstruction of model equations from time series. Phys. Rev. E **65**, 026205 (2002)
118. C. Letellier, L.A. Aguirre, J. Maquet, Relation between observability and differential embeddings for nonlinear dynamics. Phys. Rev. E **71**, 066213 (2005)
119. D.S. Broomhead, G.P. King, Extracting qualitative dynamics from experimental data. Phys. D: Nonlinear Phenom. **20**(2), 217–236 (1986)
120. H.D.I. Abarbanel, M.B. Kennel, Local false nearest neighbors and dynamical dimensions from observed chaotic data. Phys. Rev. E **47**, 3057–3068 (1993)
121. L. Cao, Practical method for determining the minimum embedding dimension of a scalar time series. Phys. D: Nonlinear Phenom. **110**(1–2), 43–50 (1997)
122. L.M. Pecora, L. Moniz, J. Nichols, T.L. Carroll, A unified approach to attractor reconstruction. Chaos: Interdiscip. J. Nonlinear Sci. **17**(1), 013110 (2007)
123. D.S. Broomhead, J.P. Huke, M.R. Muldoon, Linear filters and non-linear systems. J. R. Stat. Soc.: Ser. B (Methodol.) **54**(2), 373–382 (1992)
124. J. Theiler, S. Eubank, Don't bleach chaotic data. Chaos: Interdiscip. J. Nonlinear Sci. **3**(4), 771–782 (1993)
125. K.H. Kraemer, G. Datseris, J. Kurths, I.Z. Kiss, J.L. Ocampo-Espindola, N. Marwan, A unified and automated approach to attractor reconstruction. New J. Phys. **23**(3), 033017 (2021)
126. J. Isensee, G. Datseris, U. Parlitz, Predicting spatio-temporal time series using dimension reduced local states. J. Nonlinear Sci. **30**, 713–735 (2020)
127. C. Bandt, B. Pompe, Permutation entropy: a natural complexity measure for time series. Phys. Rev. Lett. **88**(17), 4 (2002)
128. B. Fadlallah, B. Chen, A. Keil, J. Príncipe, Weighted-permutation entropy: a complexity measure for time series incorporating amplitude information. Phys. Rev. E - Stat. Nonlinear Soft Matter Phys. **87**(2), 1–7 (2013)
129. A. Schlemmer, S. Berg, T. Lilienkamp, S. Luther, U. Parlitz, Spatiotemporal permutation entropy as a measure for complexity of cardiac arrhythmia. Front. Phys. **6**, 1–13 (2018)
130. A. Myers, F.A. Khasawneh, On the automatic parameter selection for permutation entropy. Chaos: Interdiscip. J. Nonlinear Sci. **30**(3), 033130 (2020)

131. S. Pethel, D. Hahs, Exact test of independence using mutual information. Entropy **16**(5), 2839–2849 (2014)
132. T. Schreiber, Measuring information transfer. Phys. Rev. Lett. **85**(2), 461–464 (2000)
133. M. Paluš, V. Komárek, Z. Hrnčíř, K. Štěrbová, Synchronization as adjustment of information rates: detection from bivariate time series. Phys. Rev. E **63**, 046211 (2001)
134. M. Ragwitz, H. Kantz, Markov models from data by simple nonlinear time series predictors in delay embedding spaces. Phys. Rev. E - Stat. Phys. Plasmas Fluids Relat. Interdiscip. Top. **65**(5), 12 (2002)
135. T. Bossomaier, L. Barnett, M. Harré, J.T. Lizier, *An Introduction to Transfer Entropy: Information Flow in Complex Systems* (Springer, Berlin, 2016)
136. M. Wibral, *Directed Information Measures in Neuroscience* (Springer, Heidelberg, 2014)
137. M. Staniek, K. Lehnertz, Symbolic transfer entropy. Phys. Rev. Lett. **100**, 158101 (2008)
138. A. Krakovská, J. Jakubík, M. Chvosteková, D. Coufal, N. Jajcay, M. Paluš, Comparison of six methods for the detection of causality in a bivariate time series. Phys. Rev. E **97**(4), 1–14 (2018)
139. J. Runge, S. Bathiany, E. Bollt, G. Camps-Valls, D. Coumou, E. Deyle, C. Glymour, M. Kretschmer, M.D. Mahecha, J. Muñoz-Marí, E.H. van Nes, J. Peters, R. Quax, M. Reichstein, M. Scheffer, B. Schölkopf, P. Spirtes, G. Sugihara, J. Sun, K. Zhang, J. Zscheischler, Inferring causation from time series in Earth system sciences. Nat. Commun. **10**(1), 1–13 (2019)
140. G. Sugihara, R. May, H. Ye, C.H. Hsieh, E. Deyle, M. Fogarty, S. Munch, Detecting causality in complex ecosystems. Science **338**(6106), 496–500 (2012)
141. H. Ye, E.R. Deyle, L.J. Gilarranz, G. Sugihara, Distinguishing time-delayed causal interactions using convergent cross mapping. Sci. Rep. **5**, 1–9 (2015)
142. A. Čenys, G. Lasiene, K. Pyragas, Estimation of interrelation between chaotic observables. Phys. D: Nonlinear Phenom. **52**(2), 332–337 (1991)
143. J. Arnhold, P. Grassberger, K. Lehnertz, C.E. Elger, A robust method for detecting interdependences: application to intracranially recorded EEG. Phys. D: Nonlinear Phenom. **134**(4), 419–430 (1999)
144. Y. Hirata, J.M. Amigó, Y. Matsuzaka, R. Yokota, H. Mushiake, K. Aihara, Detecting causality by combined use of multiple methods: climate and brain examples. PLOS One **11**(7), e0158572 (2016)
145. J. Theiler, S. Eubank, A. Longtin, B. Galdrikian, J.D. Farmer, Testing for nonlinearity in time series: the method of surrogate data. Phys. D: Nonlinear Phenom. **58**(1–4), 77–94 (1992)
146. G. Lancaster, D. Iatsenko, A. Pidde, V. Ticcinelli, A. Stefanovska, Surrogate data for hypothesis testing of physical systems. Phys. Rep. **748**, 1–60 (2018)
147. T. Schreiber, A. Schmitz, Surrogate time series. Phys. D: Nonlinear Phenom. **142**(3–4), 346–382 (2000)
148. T. Schreiber, A. Schmitz, Discrimination power of measures for nonlinearity in a time series. Phys. Rev. E - Stat. Phys. Plasmas Fluids Relat. Interdiscip. Top. **55**(5), 5443–5447 (1997)
149. T. Nakamura, M. Small, Y. Hirata, Testing for nonlinearity in irregular fluctuations with long-term trends. Phys. Rev. E - Stat. Nonlinear Soft Matter Phys. **74**(2), 1–8 (2006)
150. M. Tabor, *Chaos and Integrability in Nonlinear Dynamics* (Wiley, New York, 1989)
151. Y.G. Sinai, Dynamical systems with elastic reflections. Russ. Math. Surv. **25**(2), 137–189 (1970)
152. N. Chernov, Entropy, Lyapunov exponents, and mean free path for billiards. J. Stat. Phys. **88**(1–2), 1–29 (1997)
153. Wikipedia contributors, Dynamical billiards — Wikipedia, the free encyclopedia (2020). Accessed 18 Sept 2020
154. Wikipedia contributors, Hadamard's dynamical system — Wikipedia, the free encyclopedia (2020). Accessed 18 Sept 2020
155. L.A. Bunimovich, Mushrooms and other billiards with divided phase space. Chaos: Interdiscip. J. Nonlinear Sci. **11**(4), 802–808 (2001)
156. J.P. Eckmann, S.O. Kamphorst, D. Ruelle et al., Recurrence plots of dynamical systems. World Sci. Ser. Nonlinear Sci. Ser. A **16**, 441–446 (1995)

157. C.L. Webber, N. Marwan (eds.), *Recurrence Quantification Analysis* (Springer International Publishing, New York, 2015)
158. N. Marwan, M.C. Romano, M. Thiel, J. Kurths, Recurrence plots for the analysis of complex systems. Phys. Rep. **438**(5–6), 237–329 (2007)
159. N. Marwan, How to avoid potential pitfalls in recurrence plot based data analysis. Int. J. Bifurc. Chaos **21**(04), 1003–1017 (2011)
160. G. Page, C. Antoine, C.P. Dettmann, J. Talbot, The iris billiard: critical geometries for global chaos. Chaos: Interdiscip. J. Nonlinear Sci. **30**(12), 123105 (2020)
161. J.D. Meiss, Average exit time for volume-preserving maps. Chaos **7**(1), 139–147 (1997)
162. J.D. Meiss, Thirty years of turnstiles and transport. Chaos **25**(9), 097602 (2015)
163. J.C. Oxtoby, *The Poincaré Recurrence Theorem* (Springer US, New York, 1971), pp. 65–69
164. C. Ulcigrai, Poincare recurrence, online lecture notes. Accessed 3 Nov 2020
165. I. Kovacic, M.J. Brennan (eds.), *The Duffing Equation: Nonlinear Oscillators and Their Behaviour* (Wiley, New York, 2011)
166. G. Duffing, *Erzwungene Schwingungen bei veränderlicher Eigenfrequenz und ihre technische Bedeutung* (Vieweg, Braunschweig, 1918)
167. C. Scheffczyk, U. Parlitz, T. Kurz, W. Knop, W. Lauterborn, Comparison of bifurcation structures of driven dissipative nonlinear oscillators. Phys. Rev. A **43**, 6495–6502 (1991)
168. T. Kurz, W. Lauterborn, Bifurcation structure of the Toda oscillator. Phys. Rev. A **37**, 1029–1031 (1988)
169. U. Parlitz, W. Lauterborn, Superstructure in the bifurcation set of the Duffing equation $\ddot{x} + d\dot{x} + x + x^3 = f \cos(\omega t)$. Phys. Lett. A **107**(8), 351–355 (1985)
170. U. Parlitz, Common dynamical features of periodically driven strictly dissipative oscillators. Int. J. Bifurc. Chaos **03**(03), 703–715 (1993)
171. T. Uezu, Y. Aizawa, Topological character of a periodic solution in three-dimensional ordinary differential equation system. Prog. Theor. Phys. **68**(6), 1907–1916 (1982)
172. U. Parlitz, W. Lauterborn, Resonances and torsion numbers of driven dissipative nonlinear oscillators. Z. Naturforsch. A **41**(4), 605–614 (1986)
173. V. Englisch, U. Parlitz, W. Lauterborn, Comparison of winding-number sequences for symmetric and asymmetric oscillatory systems. Phys. Rev. E **92**, 022907 (2015)
174. H.G. Solari, R. Gilmore, Relative rotation rates for driven dynamical systems. Phys. Rev. A **37**, 3096–3109 (1988)
175. H.G. Solari, R. Gilmore, Organization of periodic orbits in the driven Duffing oscillator. Phys. Rev. A **38**, 1566–1572 (1988)
176. R. Gilmore, Topological analysis of chaotic dynamical systems. Rev. Mod. Phys. **70**, 1455–1529 (1998)
177. P. Beiersdorfer, J.-M. Wersinger, Y. Treve, Topology of the invariant manifolds of period-doubling attractors for some forced nonlinear oscillators. Phys. Lett. A **96**(6), 269–272 (1983)
178. R. Gilmore, M. Lefranc. *The Topology of Chaos: Alice in Stretch and Squeezeland* (Wiley, New York, 2003)
179. U. Parlitz, V. Englisch, C. Scheffczyk, W. Lauterborn, Bifurcation structure of bubble oscillators. J. Acoust. Soc. Am. **88**(2), 1061–1077 (1990)
180. W. Lauterborn, Numerical investigation of nonlinear oscillations of gas bubbles in liquids. J. Acoust. Soc. Am. **59**(2), 283–293 (1976)
181. B. van der Pol, D.Sc. Jun, Lxxxviii. on "relaxation-oscillations". Lond. Edinb. Dublin Philos. Mag. J. Sci. **2**(11), 978–992 (1926)
182. B. van der Pol, J. van der Mark, Frequency demultiplication. Nature **120**, 363–364 (1927)
183. M.L. Cartwright, J.E. Littlewood, On non-linear differential equations of the second order, i. J. Lond. Math. Soc. **20**(3), 180 (1945)
184. U. Parlitz, W. Lauterborn, Period-doubling cascades and devil's staircases of the driven van der Pol oscillator. Phys. Rev. A **36**, 1428–1434 (1987)
185. R. Mettin, U. Parlitz, W. Lauterborn, Bifurcation structure of the driven van der Pol oscillator. Int. J. Bifurc. Chaos **03**(06), 1529–1555 (1993)

186. J.-M. Ginoux, C. Letellier, Van der Pol and the history of relaxation oscillations: toward the emergence of a concept. Chaos: Interdisc. J. Nonlinear Sci. **22**(2), 023120 (2012)
187. A. Pikovsky, M. Rosenblum, J. Kurths, *Synchronization: a Universal Concept in Nonlinear Sciences*. Cambridge Nonlinear Science Series. (Cambridge University Press, Cambridge, 2001)
188. S. Strogatz, *Sync: How Order Emerges from Chaos in the Universe, Nature and Daily Life* (Hachette Books, New York, 2004)
189. A. Balanov, N. Janson, D. Postnov, O. Sosnovtseva, *Synchronization - From Simple to Complex* (Springer, Berlin, 2009)
190. H.G. Schuster (ed.), *Handbook of Chaos Control* (Wiley-VCH Verlag GmbH & Co. KGaA, New York, 1999)
191. E. Schöll, H.G. Schuster (eds.), *Handbook of Chaos Control* (Wiley, New York, 2007)
192. H. Araki (ed.), *Self-entrainment of a Population of Coupled Non-linear Oscillators* (Springer, Berlin, 1975)
193. S. Strogatz, D. Abrams, A. McRobie, B. Eckhardt, E. Ott, Crowd synchrony on the London millennium bridge. Nature **438**, 43–44 (2005)
194. B. Eckhardt, E. Ott, Crowd synchrony on the London millennium bridge. Chaos: Interdiscip. J. Nonlinear Sci. **16**(4), 041104 (2006)
195. J.A. Acebrón, L.L. Bonilla, C.J.P. Vicente, F. Ritort, R. Spigler, The Kuramoto model: a simple paradigm for synchronization phenomena. Rev. Mod. Phys. **77**, 137–185 (2005)
196. L.M. Pecora, T.L. Carroll, Synchronization in chaotic systems. Phys. Rev. Lett. **64**, 821–824 (1990)
197. L.M. Pecora, T.L. Carroll, Synchronization of chaotic systems. Chaos: Interdiscip. J. Nonlinear Sci. **25**(9), 097611 (2015)
198. D. Eroglu, J.S.W. Lamb, T. Pereira, Synchronisation of chaos and its applications. Contemp. Phys. **58**(3), 207–243 (2017)
199. A.S. Pikovsky, P. Grassberger, Symmetry breaking bifurcation for coupled chaotic attractors. J. Phys. A: Math. Gen. **24**(19), 4587–4597 (1991)
200. P. Ashwin, J. Buescu, I. Stewart, Bubbling of attractors and synchronisation of chaotic oscillators. Phys. Lett. A **193**(2), 126–139 (1994)
201. P. Ashwin, J. Buescu, I. Stewart, From attractor to chaotic saddle: a tale of transverse instability. Nonlinearity **9**(3), 703–737 (1996)
202. M.G. Rosenblum, A.S. Pikovsky, J. Kurths, Phase synchronization of chaotic oscillators. Phys. Rev. Lett. **76**, 1804–1807 (1996)
203. U. Parlitz, L. Junge, W. Lauterborn, L. Kocarev, Experimental observation of phase synchronization. Phys. Rev. E **54**, 2115–2117 (1996)
204. D. Maza, A. Vallone, H. Mancini, S. Boccaletti, Experimental phase synchronization of a chaotic convective flow. Phys. Rev. Lett. **85**, 5567–5570 (2000)
205. V.S. Afraimovich, N.N. Verichev, M.I. Rabinovich, Stochastic synchronization of oscillation in dissipative systems. Radiophys. Quantum Electron. **29**, 795–803 (1986)
206. N.F. Rulkov, M.M. Sushchik, L.S. Tsimring, H.D.I. Abarbanel, Generalized synchronization of chaos in directionally coupled chaotic systems. Phys. Rev. E **51**, 980–994 (1995)
207. H.D.I. Abarbanel, N.F. Rulkov, M.M. Sushchik, Generalized synchronization of chaos: the auxiliary system approach. Phys. Rev. E **53**, 4528–4535 (1996)
208. L. Kocarev, U. Parlitz, Generalized synchronization, predictability, and equivalence of unidirectionally coupled dynamical systems. Phys. Rev. Lett. **76**, 1816–1819 (1996)
209. U. Parlitz, L. Junge, L. Kocarev, Subharmonic entrainment of unstable period orbits and generalized synchronization. Phys. Rev. Lett. **79**, 3158–3161 (1997)
210. U. Parlitz, Detecting generalized synchronization. Nonlinear Theory Appl. **3**(2), 114–127 (2012)
211. J. Stark, Invariant graphs for forced systems. Phys. D: Nonlinear Phenom. **109**(1), 163–179 (1997) (*Proceedings of the Workshop on Physics and Dynamics Between Chaos, Order, and Noise*)

212. B.R. Hunt, E. Ott, J.A. Yorke, Differentiable generalized synchronization of chaos. Phys. Rev. E **55**, 4029–4034 (1997)
213. H. Jaeger, A tutorial on training recurrent neural networks, covering BPPT, RTRL, EKF and the "echo state network" approach. GMD Rep. **159**, 48 (2002)
214. H. Jaeger, H. Haas, Harnessing nonlinearity: predicting chaotic systems and saving energy in wireless communication. Science **304**, 78–80 (2004)
215. U. Parlitz, A. Hornstein, Dynamical prediction of chaotic time series. Chaos Complex. Lett. **1**(2), 135–144 (2005)
216. Z. Lu, B.R. Hunt, E. Ott, Attractor reconstruction by machine learning. Chaos: Interdiscip. J. Nonlinear Sci. **28**(6), 061104 (2018)
217. J. Pathak, B. Hunt, M. Girvan, Z. Lu, E. Ott, Model-free prediction of large spatiotemporally chaotic systems from data: a reservoir computing approach. Phys. Rev. Lett. **120**, 024102 (2018)
218. R.S. Zimmermann, U. Parlitz, Observing spatio-temporal dynamics of excitable media using reservoir computing. Chaos **28**(4), 043118 (2018)
219. A.-L. Barabási, R. Albert, Emergence of scaling in random networks. Science **286**(5439), 509–512 (1999)
220. D.J. Watts, S.H. Strogatz, Collective dynamics of 'small-world' networks. Nature **393**(6684), 440–442 (1998)
221. R. Albert, H. Jeong, A.-L. Barabasi, Erratum: correction: error and attack tolerance of complex networks. Nature **409**(6819), 542–542 (2001)
222. R. Albert, A.-L. Barabási, Statistical mechanics of complex networks. Rev. Mod. Phys. **74**(1), 47 (2002)
223. S.H. Strogatz, Exploring complex networks. Nature **410**(6825), 268–276 (2001)
224. M. Newman, The structure and function of complex networks. SIAM Rev. **45**, 167 (2003)
225. V. Latora, V. Nicosia, G. Russo, *Complex Networks* (Cambridge University Press, Cambridge, 2017)
226. G. Chen, X. Wang, X. Li, *Fundamentals of Complex Networks* (Wiley, Singapore, 2014)
227. P. Erdos, A. Rényi et al., On the evolution of random graphs. Publ. Math. Inst. Hung. Acad. Sci. **5**(1), 17–60 (1960)
228. S. Boccaletti, V. Latora, Y. Moreno, M. Chavez, D.-U. Hwang, Complex networks: structure and dynamics. Phys. Rep. **424**(4), 175–308 (2006)
229. A. Arenas, A. Díaz-Guilera, J. Kurths, Y. Moreno, C. Zhou, Synchronization in complex networks. Phys. Rep. **469**(3), 93–153 (2008)
230. D.M. Abrams, S.H. Strogatz, Chimera states for coupled oscillators. Phys. Rev. Lett. **93**, 174102 (2004)
231. Y. Kuramoto, D. Battogtokh, Coexistence of coherence and incoherence in nonlocally coupled phase oscillators. Nonlinear Phenom. Complex Syst. **5**(4), 380–385 (2002)
232. M. Wolfrum, O.E. Omel'chenko, Chimera states are chaotic transients. Phys. Rev. E **84**, 015201 (2011)
233. G. Filatrella, A. Nielsen, N. Pedersen, Analysis of a power grid using a Kuramoto-like model. Eur. Phys. J. B **61**, 485–491 (2008)
234. M. Rohden, A. Sorge, M. Timme, D. Witthaut, Self-organized synchronization in decentralized power grids. Phys. Rev. Lett. **109**, 064101 (2012)
235. F. Dörfler, M. Chertkov, F. Bullo, Synchronization in complex oscillator networks and smart grids. Proc. Natl. Acad. Sci. **110**(6), 2005–2010 (2013)
236. M. Anvari, F. Hellmann, X. Zhang, Introduction to focus issue: dynamics of modern power grids. Chaos: Interdiscip. J. Nonlinear Sci. **30**(6), 063140 (2020)
237. D. Braess, A. Nagurney, T. Wakolbinger, On a paradox of traffic planning. Transp. Sci. **39**(4), 446–450 (2005)
238. D. Witthaut, M. Timme, Braess's paradox in oscillator networks, desynchronization and power outage. New J. Phys. **14**(8), 083036 (2012)
239. I.Z. Kiss, J.C. Miller, P.L. Simon, *Mathematics of Epidemics on Networks* (Springer, Cham, 2017)

240. E. Bonabeau, Agent-based modeling: methods and techniques for simulating human systems. Proc. Natl. Acad. Sci. **99**(Supplement 3), 7280–7287 (2002)
241. L. Hamill, N. Gilbert, *Agent-Based Modelling in Economics* (Wiley, New York, 2015)
242. M.C. Cross, P.C. Hohenberg, Pattern formation outside of equilibrium. Rev. Mod. Phys. **65**, 851–1112 (1993)
243. M. Cross, H. Greenside, *Pattern Formation and Dynamics in Nonequilibrium Systems* (Cambridge University Press, Cambridge, 2009)
244. J.D. Murray, *Mathematical Biology* (Springer, Berlin, 1989)
245. A.M. Turing, The chemical basis of morphogenesis. Philos. Trans. R. Soc. Lond. Ser. B Biol. Sci. **237**(641), 37–72 (1952)
246. L. Yang, I.R. Epstein, Oscillatory Turing patterns in reaction-diffusion systems with two coupled layers. Phys. Rev. Lett. **90**, 178303 (2003)
247. A.M. Zhabotinsky, A.N. Zaikin, Autowave processes in a distributed chemical system. J. Theor. Biol. **40**(1), 45–61 (1973)
248. A.M. Zhabotinsky, A history of chemical oscillations and waves. Chaos: Interdiscip. J. Nonlinear Sci. **1**(4), 379–386 (1991)
249. J.J. Tyson, K.A. Alexander, V.S. Manoranjan, J.D. Murray, Spiral waves of cyclic AMP in a model of slime mold aggregation. Phys. D: Nonlinear Phenom. **34**(1), 193–207 (1989)
250. S. Jakubith, H.H. Rotermund, W. Engel, A. von Oertzen, G. Ertl, Spatiotemporal concentration patterns in a surface reaction: propagating and standing waves, rotating spirals, and turbulence. Phys. Rev. Lett. **65**, 3013–3016 (1990)
251. V.I. Krinsky, Spread of excitation in an inhomogeneous medium (state similar to cardiac fibrillation). Biofizika **11**, 676–683 (1066)
252. A.T. Winfree, Electrical turbulence in three-dimensional heart muscle. Science **266**(5187), 1003–1006 (1994)
253. F.H. Fenton, E.M. Cherry, Models of cardiac cell. Scholarpedia **3**(8), 1868 (2008). Revision #91508
254. A.T. Winfree, Electrical instability in cardiac muscle: phase singularities and rotors. J. Theor. Biol. **138**(3), 353–405 (1989)
255. R.A. Gray, A.M. Pertsov, J. Jalife, Spatial and temporal organization during cardiac fibrillation. Nature **392**(6671), 75–78 (1998)
256. A.N. Iyer, R.A. Gray, An experimentalist's approach to accurate localization of phase singularities during reentry. Ann. Biomed. Eng. **29**(1), 47–59 (2001)
257. D. Barkley, M. Kness, L.S. Tuckerman, Spiral-wave dynamics in a simple model of excitable media: the transition from simple to compound rotation. Phys. Rev. A **42**, 2489–2492 (1990)
258. A.V. Holden, H. Zhang, Modelling propagation and re-entry in anisotropic and smoothly heterogeneous cardiac tissue. J. Chem. Soc. Faraday Trans. **89**, 2833–2837 (1993)
259. F. Fenton, A. Karma, Vortex dynamics in three-dimensional continuous myocardium with fiber rotation: filament instability and fibrillation. Chaos: Interdiscip. J. Nonlinear Sci. **8**(1), 20–47 (1998)
260. C.D. Marcotte, R.O. Grigoriev, Dynamical mechanism of atrial fibrillation: a topological approach. Chaos: Interdiscip. J. Nonlinear Sci. **27**(9), 093936 (2017)
261. D.R. Gurevich, R.O. Grigoriev, Robust approach for rotor mapping in cardiac tissue. Chaos: Interdiscip. J. Nonlinear Sci. **29**(5), 053101 (2019)
262. R.H. Clayton, A.V. Holden, Dynamics and interaction of filaments in a computational model of re-entrant ventricular fibrillation. Phys. Med. Biol. **47**(10), 1777–1792 (2002)
263. R.M. Zaritski, S.F. Mironov, A.M. Pertsov, Intermittent self-organization of scroll wave turbulence in three-dimensional excitable media. Phys. Rev. Lett. **92**, 168302 (2004)
264. F. Spreckelsen, D. Hornung, O. Steinbock, U. Parlitz, S. Luther, Stabilization of three-dimensional scroll waves and suppression of spatiotemporal chaos by heterogeneities. Phys. Rev. E **92**, 042920 (2015)
265. J. Christoph, M. Chebbok, C. Richter, J. Schröder-Schetelig, P. Bittihn, S. Stein, I. Uzelac, F.H. Fenton, G. Hasenfuß, R.F. Gilmour Jr., S. Luther, Electromechanical vortex filaments during cardiac fibrillation. Nature **555**, 667–672 (2018)

266. Y.-C. Lai, T. Tél, *Transient Chaos* (Springer, New York, 2011)
267. Z. Qu, Critical mass hypothesis revisited: role of dynamical wave stability in spontaneous termination of cardiac fibrillation. Am. J. Physiol.-Heart Circ. Physiol. **290**(1), H255–H263 (2006). PMID: 16113075
268. T. Lilienkamp, J. Christoph, U. Parlitz, Features of chaotic transients in excitable media governed by spiral and scroll waves. Phys. Rev. Lett. **119**(5), 054101 (2017)
269. T. Lilienkamp, U. Parlitz, Terminal transient phase of chaotic transients. Phys. Rev. Lett. **120**(9), 094101 (2018)
270. S. Luther, F.H. Fenton, B.G. Kornreich, A. Squires, P. Bittihn, D. Hornung, M. Zabel, J. Flanders, A. Gladuli, L. Campoy, E.M. Cherry, G. Luther, G. Hasenfuss, V.I. Krinsky, A. Pumir, R.F. Gilmour, E. Bodenschatz, Low-energy control of electrical turbulence in the heart. Nature **475**, 235–239 (2011)
271. Y. Kuramoto, Diffusion-induced chaos in reaction systems. Prog. Theor. Phys. Suppl. **64**, 346–367 (1978)
272. G. Sivashinsky, On flame propagation under conditions of stoichiometry. SIAM J. Appl. Math. **39**(1), 67–82 (1980)
273. G.I. Sivashinsky, Nonlinear analysis of hydrodynamic instability in laminar flames—I. Derivation of basic equations, in *Dynamics of Curved Fronts*, ed. by P. Pelcé (Academic, San Diego, 1988), pp. 459–488
274. J.M. Hyman, B. Nicolaenko, The Kuramoto-Sivashinsky equation: a bridge between PDE's and dynamical systems. Physica D: Nonlinear Phenom. **18**(1), 113–126 (1986)
275. D.M. Raup, J.J. Sepkoski, Mass extinctions in the marine fossil record. Science **215**(4539), 1501–1503 (1982)
276. J. Alroy, Dynamics of origination and extinction in the marine fossil record. Proc. Natl. Acad. Sci. **105**(Supplement 1), 11536–11542 (2008)
277. Wikipedia contributors, Extinction event — Wikipedia, the free encyclopedia (2021). Accessed 1 April 2021
278. D.H. Rothman, Characteristic disruptions of an excitable carbon cycle. Proc. Natl. Acad. Sci. USA **116**(30), 14813–14822 (2019)
279. D.H. Rothman, Earth's carbon cycle: a mathematical perspective. Bull. Am. Math. Soc. **52**(1), 47–64 (2014)
280. D.H. Rothman, Thresholds of catastrophe in the Earth system. Sci. Adv. **3**(9), 1–13 (2017)
281. D.H. Rothman, Atmospheric carbon dioxide levels for the last 500 million years. Proc. Natl. Acad. Sci. USA **99**(7), 4167–4171 (2002)
282. D.H. Rothman, J.M. Hayes, R.E. Summons, Dynamics of the Neoproterozoic carbon cycle. Proc. Natl. Acad. Sci. USA **100**(14), 8124–8129 (2003)
283. H.G. Marshall, J.C.G. Walker, W.R. Kuhn, Long-term climate change and the geochemical cycle of carbon. J. Geophys. Res. **93**(D1), 791 (1988)
284. T.J. Crowley, G.R. North, Abrupt climate change and extinction events in Earth history. Science **240**(4855), 996–1002 (1988)
285. M. Crucifix, Oscillators and relaxation phenomena in Pleistocene climate theory. Philos. Trans. R. Soc. A: Math. Phys. Eng. Sci. **370**(1962), 1140–1165 (2012)
286. B. de Saedeleer, M. Crucifix, S. Wieczorek, Is the astronomical forcing a reliable and unique pacemaker for climate? A conceptual model study. Clim. Dyn. **40**(1–2), 273–294 (2013)
287. S. Wieczorek, P. Ashwin, C.M. Luke, P.M. Cox, Excitability in ramped systems: the compost-bomb instability. Proc. R. Soc. A: Math. Phys. Eng. Sci. **467**(2129), 1243–1269 (2010)
288. C. Nicolis, G. Nicolis, Is there a climatic attractor? Nature **311**(5986), 529–532 (1984)
289. K. Fraedrich, Estimating the dimensions of weather and climate attractors. J. Atmos. Sci. **43**(5), 419–432 (1986)
290. C. Nicolis, G. Nicolis, Evidence for climatic attractors. Nature **326**(6112), 523–523 (1987)
291. K. Fraedrich, Estimating weather and climate predictability on attractors. J. Atmos. Sci. **44**(4), 722–728 (1987)
292. C. Essex, T. Lookman, M.A.H. Nerenberg, The climate attractor over short timescales. Nature **326**(6108), 64–66 (1987)

293. A.A. Tsonis, J.B. Elsner, The weather attractor over very short timescales. Nature **333**(6173), 545–547 (1988)

294. K. Fraedrich, L.M. Leslie, Estimates of cyclone track predictability. I: tropical cyclones in the Australian region. Q. J. R. Meteorol. Soc. **115**(485), 79–92 (1989)

295. C.L. Keppenne, C. Nicolis, Global properties and local structure of the weather attractor over Western Europe. J. Atmos. Sci. **46**(15), 2356–2370 (1989)

296. M.B. Sharifi, K.P. Georgakakos, I. Rodriguez-Iturbe, Evidence of deterministic chaos in the pulse of storm rainfall. J. Atmos. Sci. **47**(7), 888–893 (1990)

297. K. Fraedrich, R. Wang, Estimating the correlation dimension of an attractor from noisy and small datasets based on re-embedding. Phys. D: Nonlinear Phenom. **65**(4), 373–398 (1993)

298. P. Grassberger, Do climatic attractors exist? Nature **323**(6089), 609–612 (1986)

299. P. Grassberger, Evidence for climatic attractors. Nature **326**(6112), 524–524 (1987)

300. I. Procaccia, Complex or just complicated? Nature **333**(6173), 498–499 (1988)

301. E.N. Lorenz, Dimension of weather and climate attractors. Nature **353**(6341), 241–244 (1991)

302. J.R. Petit, J. Jouzel, D. Raynaud, N.I. Barkov, J.-M. Barnola, I. Basile, M. Bender, J. Chappellaz, M. Davis, G. Delaygue, M. Delmotte, V.M. Kotlyakov, M. Legrand, V.Y. Lipenkov, C. Lorius, L. PÉpin, C. Ritz, E. Saltzman, M. Stievenard, Climate and atmospheric history of the past 420, 000 years from the Vostok ice core, Antarctica. Nature **399**(6735), 429–436 (1999)

303. M.M. Milankovitch, *Canon of Insolation and the Iceage Problem*, vol. 132 (Koniglich Serbische Akademice Beograd Special Publication, 1941)

304. B. Saltzman, A.R. Hansen, K.A. Maasch, The late quaternary glaciations as the response of a three-component feedback system to Earth-orbital forcing. J. Atmos. Sci. **41**(23), 3380–3389 (1984)

305. B. Saltzman, *Dynamical Paleoclimatology: Generalized Theory of Global Climate Change* (Elsevier, Amsterdam, 2001)

306. B. Saltzman, K.A. Maasch, A first-order global model of late Cenozoic climatic change II. Further analysis based on a simplification of CO2 dynamics. Clim. Dyn. **5**(4), 201–210 (1991)

307. H.A. Dijkstra, *Nonlinear Climate Dynamics* (Cambridge University Press, Cambridge, 2009)

308. H. Kaper, H. Engler, *Mathematics and Climate* (Society for Industrial and Applied Mathematics, Philadelphia, 2013)

309. T. Palmer, R. Hagedorn (eds.), *Predictability of Weather and Climate* (Cambridge University Press, Cambridge, 2006)

310. E.N. Lorenz, The predictability of a flow which possesses many scales of motion. Tellus **21**(3), 289–307 (1969)

311. T. Thornes, P. Düben, T. Palmer, On the use of scale-dependent precision in Earth system modelling. Q. J. R. Meteorol. Soc. **143**(703), 897–908 (2017)

312. J. Brisch, H. Kantz, Power law error growth in multi-hierarchical chaotic systems—a dynamical mechanism for finite prediction horizon. New J. Phys. **21**(9), 093002 (2019)

313. R.M. May, Thresholds and breakpoints in ecosystems with a multiplicity of stable states. Nature **269**(5628), 471–477 (1977)

314. M. Scheffer, S. Carpenter, J.A. Foley, C. Folke, B. Walker, Catastrophic shifts in ecosystems. Nature **413**(6856), 591–596 (2001)

315. C. Folke, S. Carpenter, B. Walker, M. Scheffer, T. Elmqvist, L. Gunderson, C.S. Holling, Regime shifts, resilience, and biodiversity in ecosystem management. Annu. Rev. Ecol. Evol. Syst. **35**, 557–581 (2004)

316. C. Folke, Resilience: the emergence of a perspective for social–ecological systems analyses. Glob. Environ. Chang. **16**(3), 253–267 (2006). *Resilience, Vulnerability, and Adaptation: a Cross-Cutting Theme of the International Human Dimensions Programme on Global Environmental Change*

317. J.M. Drake, B.D. Griffen, Early warning signals of extinction in deteriorating environments. Nature **467**(7314), 456–459 (2010)

318. Z. Mukandavire, S. Liao, J. Wang, H. Gaff, D.L. Smith, J.G. Morris, Estimating the reproductive numbers for the 2008–2009 cholera outbreaks in Zimbabwe. Proc. Natl. Acad. Sci. **108**(21), 8767–8772 (2011)

319. L. Li, C.-H. Wang, S.-F. Wang, M.-T. Li, L. Yakob, B. Cazelles, Z. Jin, W.-Y. Zhang, Hemorrhagic fever with renal syndrome in china: mechanisms on two distinct annual peaks and control measures. Int. J. Biomath. **11**(02), 1850030 (2018)

320. D. Alonso, A. Dobson, M. Pascual, Critical transitions in malaria transmission models are consistently generated by superinfection. Philos. Trans. R. Soc. B: Biol. Sci. **374**(1775), 20180275 (2019)

321. R.B. Alley, Abrupt climate change. Science **299**(5615), 2005–2010 (2003)

322. T.M. Lenton, H. Held, E. Kriegler, J.W. Hall, W. Lucht, S. Rahmstorf, H.J. Schellnhuber, Tipping elements in the Earth's climate system. Proc. Natl. Acad. Sci. **105**(6), 1786–1793 (2008)

323. M. Scheffer, S.R. Carpenter, T.M. Lenton, J. Bascompte, W. Brock, V. Dakos, J. Van De Koppel, I.A. Van De Leemput, S.A. Levin, E.H. Van Nes, M. Pascual, J. Vandermeer, Anticipating critical transitions. Science **338**(6105), 344–348 (2012)

324. W. Steffen, J. Rockström, K. Richardson, T.M. Lenton, C. Folke, D. Liverman, C.P. Summerhayes, A.D. Barnosky, S.E. Cornell, M. Crucifix, J.F. Donges, I. Fetzer, S.J. Lade, M. Scheffer, R. Winkelmann, H.J. Schellnhuber, Trajectories of the Earth system in the Anthropocene. Proc. Natl. Acad. Sci. USA **115**(33), 8252–8259 (2018)

325. M. Wiedermann, R. Winkelmann, J.F. Donges, C. Eder, J. Heitzig, A. Katsanidou, E.K. Smith, Domino effects in the earth system – The potential role of wanted tipping points (2019), arXiv, pp. 1–6

326. V.I. Arnol'd (ed.), *Dynamical Systems V: Bifurcation Theory and Catastrophe Theory* (Springer, Berlin, 1994)

327. P. Ashwin, S. Wieczorek, R. Vitolo, P. Cox, Tipping points in open systems: bifurcation, noise-induced and rate-dependent examples in the climate system. Philos. Trans. R. Soc. A: Math. Phys. Eng. Sci. **370**(1962), 1166–1184 (2012)

328. P. Ashwin, A.S. von der Heydt, Extreme sensitivity and climate tipping points. J. Stat. Phys. **179**(5–6), 1531–1552 (2020)

329. H. Alkhayuon, R.C. Tyson, S. Wieczorek, Phase-sensitive tipping: how cyclic ecosystems respond to contemporary climate (2021)

330. A.N. Pisarchik, U. Feudel, Control of multistability. Phys. Rep. **540**(4), 167–218 (2014)

331. U. Feudel, A.N. Pisarchik, K. Showalter, Multistability and tipping: from mathematics and physics to climate and brain - Minireview and preface to the focus issue. Chaos **28**(3), 033501 (2018)

332. M. Scheffer, J. Bascompte, W.A. Brock, V. Brovkin, S.R. Carpenter, V. Dakos, H. Held, E.H. Van Nes, M. Rietkerk, G. Sugihara, Early-warning signals for critical transitions. Nature **461**(7260), 53–59 (2009)

333. V. Brovkin, E. Brook, J.W. Williams, S. Bathiany, T.M. Lenton, M. Barton, R.M. DeConto, J.F. Donges, A. Ganopolski, J. McManus, S. Praetorius, A. de Vernal, A. Abe-Ouchi, H. Cheng, M. Claussen, M. Crucifix, G. Gallopín, V. Iglesias, D.S. Kaufman, T. Kleinen, F. Lambert, S. van der Leeuw, H. Liddy, M.-F. Loutre, D. McGee, K. Rehfeld, R. Rhodes, A.W.R. Seddon, M.H. Trauth, L. Vanderveken, Z. Yu, Past abrupt changes, tipping points and cascading impacts in the Earth system. Nat. Geosci. **14**, 550–558 (2021)

334. V. Dakos, M. Scheffer, E.H. van Nes, V. Brovkin, V. Petoukhov, H. Held, Slowing down as an early warning signal for abrupt climate change. Proc. Natl. Acad. Sci. **105**(38), 14308–14312 (2008)

335. C. Kuehn, A mathematical framework for critical transitions: bifurcations, fast-slow systems and stochastic dynamics. Phys. D: Nonlinear Phenom. **240**(12), 1020–1035 (2011)

336. J.M.T. Thompson, J. Sieber, Predicting climate tipping as a noisy bifurcation: a review. Int. J. Bifurc. Chaos **21**(02), 399–423 (2011)

337. P. Ritchie, J. Sieber, Early-warning indicators for rate-induced tipping. Chaos **26**(9) (2016)

338. T. Wilkat, T. Rings, K. Lehnertz, No evidence for critical slowing down prior to human epileptic seizures. Chaos: Interdisc. J. Nonlinear Sci. **29**(9), 091104 (2019)

339. P.J. Menck, J. Heitzig, N. Marwan, J. Kurths, How basin stability complements the linear-stability paradigm. Nat. Phys. **9**(2), 89–92 (2013)

340. N. Wunderling, M. Gelbrecht, R. Winkelmann, J. Kurths, J.F. Donges, Basin stability and limit cycles in a conceptual model for climate tipping cascades. New J. Phys. **22**(12), 123031 (2020)

341. B. Kaszás, U. Feudel, T. Tél, Tipping phenomena in typical dynamical systems subjected to parameter drift. Sci. Rep. **9**(1), 1–12 (2019)

342. M. Ghil, V. Lucarini, The physics of climate variability and climate change. Rev. Mod. Phys. **92**(3), 35002 (2020)

343. M. Ghil, Energy-balance models: an introduction, in *Climatic Variations and Variability: Facts and Theories* (Springer Netherlands, Dordrecht, 1981), pp. 461–480

344. R. Benzi, A. Sutera, A. Vulpiani, The mechanism of stochastic resonance. J. Phys. A: Math. Gen. **14**(11), L453–L457 (1981)

345. R. Benzi, G. Parisi, A. Sutera, A. Vulpiani, Stochastic resonance in climatic change. Tellus **34**(1), 10–15 (1982)

346. L. Gammaitoni, P. Hänggi, P. Jung, F. Marchesoni, Stochastic resonance. Rev. Mod. Phys. **70**, 223–287 (1998)

347. C. Rouvas-Nicolis, G. Nicolis, Stochastic resonance. Scholarpedia **2**(11), 1474 (2007)

348. H. Stommel, Thermohaline convection with two stable regimes of flow. Tellus **13**(2), 224–230 (1961)

349. J. Lohmann, D. Castellana, P.D. Ditlevsen, H.A. Dijkstra, Abrupt climate change as rate-dependent cascading tipping point. Earth Syst. Dyn. Discuss. **2021**, 1–25 (2021)

350. M.L. Rosenzweig, R.H. MacArthur, Graphical representation and stability conditions of predator-prey interactions. Am. Nat. **97**(895), 209–223 (1963)

351. A. Vanselow, S. Wieczorek, U. Feudel, When very slow is too fast - collapse of a predator-prey system. J. Theor. Biol. **479**, 64–72 (2019)

352. V.I. Arnold, *Ordinary Differential Equations*, (Springer-Verlag Berlin Heidelberg, 1991)

353. V.I. Arnold, *Geometrical Methods in the Theory of Ordinary Differential Equations*, (Springer, New York, 1988)

354. P. Grassberger, T. Schreiber, C. Schaffrath, Nonlinear time sequence analysis, Int. J. Bifurc. Chaos **1**(3), 521–547 (1991)

355. L. Faes, A. Porta, G. Nollo, M. Javorka, Information decomposition in multivariate systems: definitions, implementation and application to cardiovascular networks. Entropy **19**(1), 5 (2017)

356. A. Kraskov, H. Stögbauer, P. Grassberger, Estimating mutual information. Phys. Rev. E **69**(6), 066138 (2004)

357. D. Witthaut, F. Hellmann, J. Kurths, S. Kettemann, H. Meyer-Ortmanns, and M. Timme, Collective nonlinear dynamics and self-organization in decentralized power grids, accepted for publication in Rev. Mod. Phys. (2022)

# Index

© The Editor(s) (if applicable) and The Author(s), under exclusive license to Springer
Nature Switzerland AG 2022
G. Datseris and U. Parlitz, *Nonlinear Dynamics*, Undergraduate Lecture Notes in Physics,
https://doi.org/10.1007/978-3-030-91032-7

Printed in the United States
by Baker & Taylor Publisher Services